ELEMENTS DETERMINED BY ATOMIC ABSORPTION

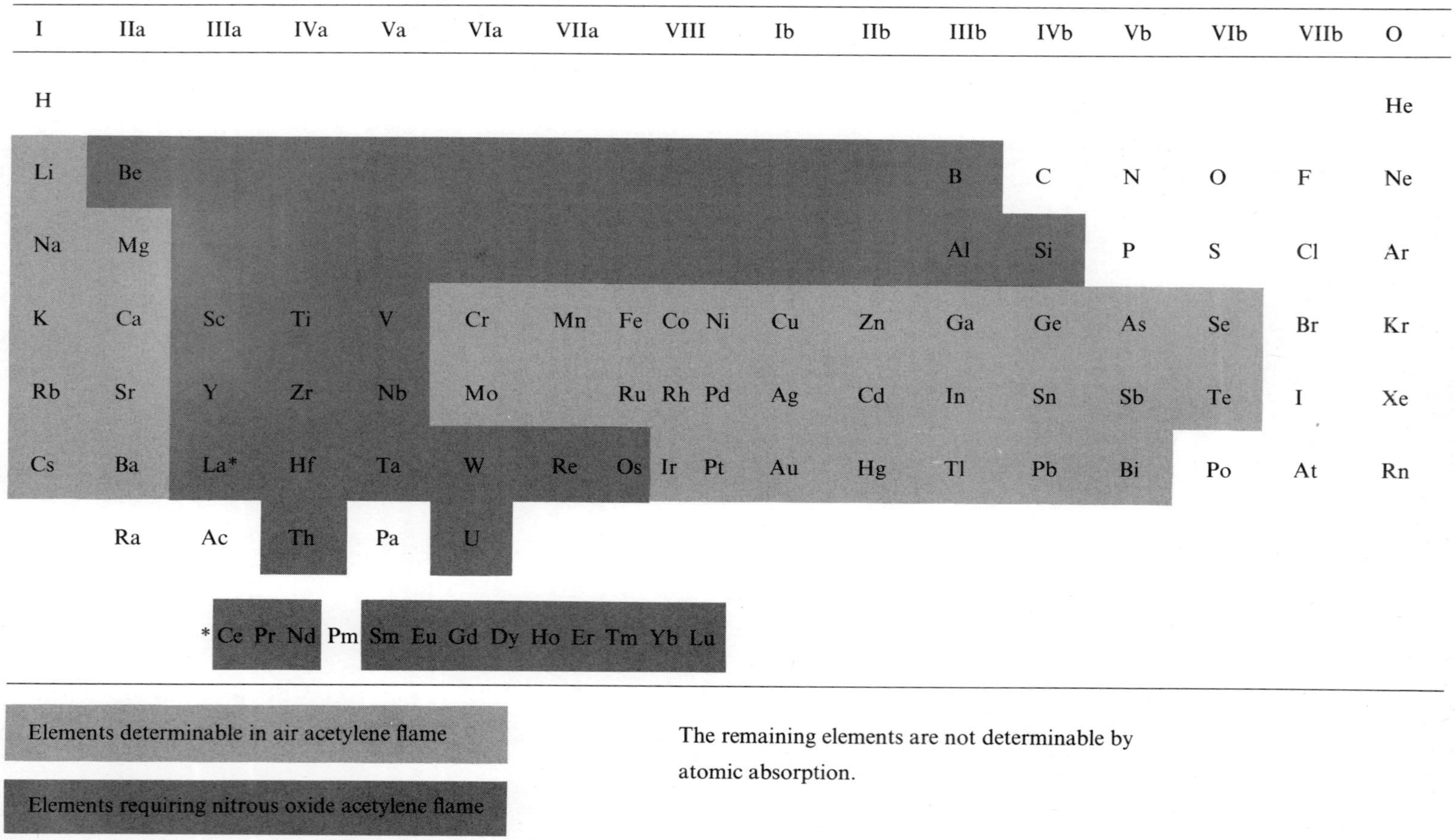

I	IIa	IIIa	IVa	Va	VIa	VIIa	VIII	Ib	IIb	IIIb	IVb	Vb	VIb	VIIb	O
H															He
Li	Be									B	C	N	O	F	Ne
Na	Mg									Al	Si	P	S	Cl	Ar
K	Ca	Sc	Ti	V	Cr	Mn	Fe Co Ni	Cu	Zn	Ga	Ge	As	Se	Br	Kr
Rb	Sr	Y	Zr	Nb	Mo		Ru Rh Pd	Ag	Cd	In	Sn	Sb	Te	I	Xe
Cs	Ba	La*	Hf	Ta	W	Re	Os Ir Pt	Au	Hg	Tl	Pb	Bi	Po	At	Rn
	Ra	Ac	Th	Pa	U										

* Ce Pr Nd Pm Sm Eu Gd Dy Ho Er Tm Yb Lu

Elements determinable in air acetylene flame

Elements requiring nitrous oxide acetylene flame

The remaining elements are not determinable by atomic absorption.

Analytical Atomic Absorption Spectrometry

Analytical Atomic Absorption Spectrometry

W. J. Price

Pye Unicam Ltd., Cambridge

HEYDEN & SON LTD

London · New York · Rheine

Heyden & Son Ltd., Spectrum House, Alderton Crescent, London NW4 3XX
Heyden & Son Inc., 225 Park Avenue, New York 10017, U.S.A.
Heyden & Son GmbH, 4440 Rheine/Westf., Münsterstrasse 22, Germany

Second printing with corrections 1974

Library of Congress Catalog Card No. 72-76558

ISBN 0 85501 045 2

Printed in Great Britain by J. W. Arrowsmith Ltd., Bristol BS3 2NT

To:

Len R. Morris and Paul Johns, for many helpful discussions, valuable suggestions and assistance,

Professor Tom West, who kindly and so readily agreed to write the Foreword,

Barbara, my wife, who not only displayed exemplary patience throughout, but prepared the bibliography and gave considerable help with the proofreading,

I tender my sincerest thanks and dedicate this book.

Foreword

More than any other technique in spectroscopy, atomic absorption spectrometry has caught the imagination of analytical chemists during recent years. Its evolution and its acceptance in all spheres of analysis involving trace metals has revolutionized analytical practice. Techniques such as solution absorptiometry and polarography have been rendered almost obsolete for trace metal analysis, and it is only in the analysis of non-metals that these time-honoured procedures can still compete.

Atomic absorption spectrometry has been widely practised now for over ten years, and, as the author indicates in his preface, there are many texts on this subject, some of which are excellent. The reader will find, however, that the author has by no means produced yet another manuscript on the same plane as the others. He has not produced a literature survey or a laboratory manual of instruction showing how elements A–Z can be determined in pure solution under specified conditions of wavelength, slit-width, etc., nor has he produced a highly theoretical account for those who are inclined to regard such a treatment essential above all else. Instead we find a carefully thought-out monographic exposition of the basic theory and practice of atomic spectrometry in flame media in relation to analysis. In his treatment the author has thought forward as well as backwards in time. The reader will find projection of trends written into the text, e.g. the relationship of fluorescence to absorption, non-flame cells in relation to flame-based techniques, and so on. This blend is very attractive indeed and it is combined with a wealth of practical know-how and personal expertise which the author has built up over a period of years. Working as he does with one of the world's leading manufacturers of spectroscopic equipment he has been uniquely situated to realize the merits and shortcomings of atomic techniques in theory and practice *vis-à-vis* other techniques both of trace and of major component metal analysis. He has been well positioned also to appreciate the needs and difficulties experienced by users of atomic absorption spectrometry in all walks of scientific investigation and to know what is most required by them. Additionally he has been a most active member of the Atomic Spectroscopy Group of the Society for Analytical Chemistry and has been associated with several other spectroscopic groups, e.g. in the Institute of Physics and the Institute of Petroleum.

All of this comes through in these pages. The text is essentially practical in nature and is very easy to read. The author is constantly looking forward and assessing the present position against probable developments. But, at the same time, there is a very thorough grasp of the fundamentals that do not change and of the practical considerations that have not been dealt with very well elsewhere in the literature.

I have known and respected the author now for several years and it does not surprise me to find that he has written such a fine monograph on this most fascinating of all the techniques of metal analysis. The reader will find that the author's enthusiasm is infectious.

Chemistry Department,
Imperial College,
London, SW7 2AY.

T. S. WEST
Professor of Analytical Chemistry

Preface

As there already exist ten or so books on atomic absorption spectroscopy what purpose can be served by yet another? This book is written specifically for the analytical chemist at his bench or the student of analytical chemistry, who now finds it difficult to select between the many procedures published and referenced in other texts, and who may consider the numerous sources of theory and background information to be confusing rather than enlightening.

The time has now come when it is not only possible but necessary to treat the atomic absorption literature selectively, and even cautiously. A textbook on practical atomic absorption must be considerably more than a guide to the literature – it must itself in principle be complete.

We attempt here to rationalize the theoretical background necessary for an adequate understanding of the subject, in both its physical and its chemical aspects. The physics is largely that of atomic spectra, while the chemistry embraces flame processes in relation to interferences as well as sample preparation.

The material has been arranged in the way in which we believe the analyst would wish to develop a knowledge of the subject. First a foundation of the basic principles is provided in which the most relevant parts of general theory are developed in the direction of the practical technique. This is followed by the instrumental interpretation of these principles – the sample handling, optical and electronic devices which allow the principles to be fully exploited. Next, analytical considerations are discussed in general terms – the way in which the equipment is made to render the best results – and the processes militating against them.

Finally the actual application of the technique is examined in some detail over a wide range of analytical topics, many of which have been investigated in the author's laboratory, and, where a direct reference is not given to a quoted analytical method, this is in fact where it originates.

After the first few hectic years in the life of this technique, which must have been tried in nearly every possible analytical situation, the preferred types of method are coming to be recognized.

Standardization, not only of the methods but also of the language of atomic absorption, is beginning to take place. Ten years is perhaps not a long time for a vocabulary to become world-wide, but with the present rapidity of literature interchange it is too long if misunderstandings remain because of a lack of common terminology.

In this book the set of definitions lately agreed by several international bodies are summarized and (it is hoped and intended) used throughout. Other, older, conventions adhered to are that (i) if acid strengths are not given, the concentrated acid that comes in the bottles is to be understood (e.g. hydrochloric acid sp.gr. 1·18, sulphuric acid sp.gr. 1·98, nitric acid sp.gr. 1.42 and perchloric acid sp.gr. 1·54), (ii) dilutions are usually quoted as, for example, '1 + 4' to avoid the ambiguity of '1 to 5' or 'five times' (somehow, a *large* dilution expressed as 'one hundred times' does *not* seem ambiguous!), (iii) the words 'precision' and 'accuracy' are used intentionally to convey the idea of reproducibility (low random error) and overall accuracy (freedom from both bias and random error) respectively.

I gratefully acknowledge permission to reproduce figures and other material from:

Dr. Alan Walsh, CSIRO, Melbourne, Australia (information in Fig. 1)
Mr. Colin Watson, Messrs Hopkin & Williams, Ltd., London (Table 7)
Dr. G. F. Kirkbright and Maxwell Scientific Inc., New York (Figs. 16, 17)
Pye Unicam Ltd., Cambridge (Figs. 9, 10)

and, in particular, I thank Dr. Lyndon Davies and the Management of Pye Unicam Ltd., Cambridge for encouragement in this undertaking.

W. JOHN PRICE
Cambridge, February 1972

Contents

Chapter 1
Introduction

DEFINITION AND HISTORICAL

Atomic Absorption Spectrometry is an analytical method for the determination of elements, based upon the absorption of radiation by free atoms. Atomic fluorescence spectrometry also enables elements to be determined but on the basis of the re-emission of the radiant energy absorbed by the free atoms.

Interaction of atoms with various forms of energy results in three very closely related spectroscopic phenomena which are commonly used in analytical laboratories – emission, absorption and fluorescence.

The extreme value of atomic absorption as a versatile laboratory technique – presently overshadowing both flame emission and fluorescence, which appear to have greater limitations – is evidenced by the almost exponential increase in the published literature on the subject during the last ten years. A further indication is the number of instruments for which licences were granted during the same period (Fig. 1). With literally just a few instruments in the world in 1960, there can, in 1972, have been no fewer than 20 000.

Atomic fluorescence shows few signs as yet of being accepted to this extent, but its possibilities in the field of simultaneous multi-element analysis are now being explored, and this, apart from high sensitivity analyses of a limited number of elements, may be its major use.

No chronicler of this subject fails to point out that the first atomic absorption observations were made in 1802 by Wollaston[453] – these were of the Fraunhofer lines in the solar spectrum – or that Kirchhoff and Bunsen (1860)[204] were first to demonstrate that atomic spectra, used in emission or absorption, could form the basis of a new and highly specific method of analysis. Emission techniques, perhaps because the spectra are more tangible and easier to measure, have become widely used. Emission instrumentation varies in sophistication from simple filter flame photometers to the point where, even in 1950, twenty or thirty elements could be determined in one sample almost in the same number of seconds.

Apart from two specialized applications, the identification of some elements in stellar atmospheres and the detection of mercury vapour in laboratory atmospheres, the real potential of atomic absorption was not realized until Walsh's exposition of the subject[415] in 1955. After a few more years of general unconcern, progress became meteoric. Improved instrumentation, more reliable sources of

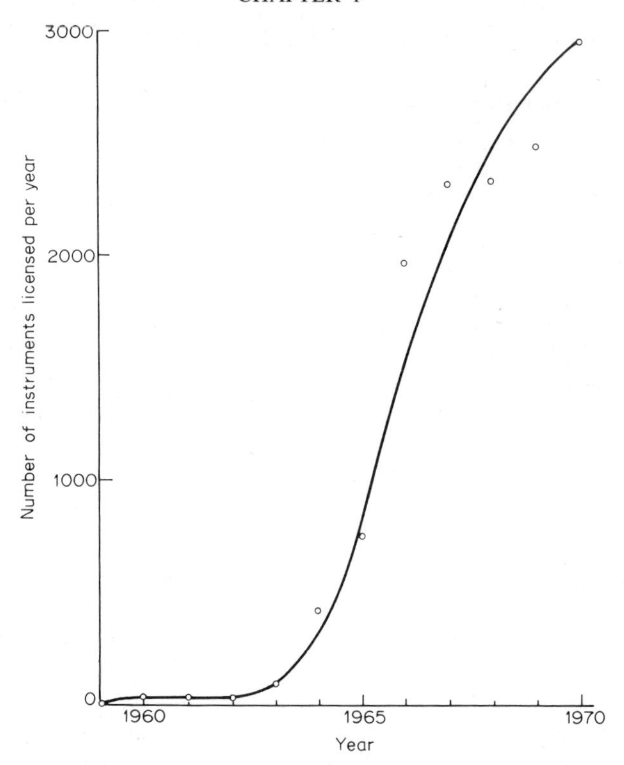

FIG. 1. Annual licensed sales of atomic absorption spectrometers 1959–1970. (Information given by CSIRO to licensees.)

resonance radiation, and hotter, more controllable flames have enabled the technique to be extended to nearly every metallic element in the periodic table. What, less than ten years ago, was thought of as an intriguing method for determining some trace elements is now used in all concentration ranges with accuracies that compare well with most other accepted techniques.

LITERATURE SOURCES

Several general and topic-orientated textbooks have already been written on atomic absorption spectroscopy.[1–11] The subject is also dealt with in at least one general textbook on spectroscopy.[12] These are listed in the general bibliography.

The present book is written particularly with the practising analytical chemist in mind. Thus, as a considerable part of the published literature is by now either outdated or simply not directly applicable, the bibliography does not pretend to be complete.

Full lists of published papers are compiled regularly elsewhere.[13] Bibliographies are also usually included among the atomic absorption instrument manufacturers' literature, and some of the manufacturers include reviews and original papers on the subject in their house journals.[13–15] Apart from the usual sources of abstracts, *Analytical Abstracts*,[16] *Chemical Abstracts*[17] and *Spectro-*

chemical Abstracts,[18] there is also a classified source, *Atomic Absorption and Flame Emission Spectroscopy Abstracts*,[19] which also includes atomic fluorescence. This is published bimonthly and aims at complete coverage, at least from 1955, of all possible journals for articles and papers on atomic flame emission, absorption and fluorescence. Reviews appear at regular intervals in the analytical and spectrochemical journals. Some of the more notable critical reviews are included in the general bibliography.[20–25] Biennial literature reviews are included in one or other sections of the annual reviews in *Analytical Chemistry*,[26] and the Society for Analytical Chemistry is publishing its *Annual Reviews in Analytical Atomic Spectroscopy*[27] from early in 1972. Apart from the general sources listed above, an atomic fluorescence bibliography has been published commercially.[28]

In the early 1960s, the novelty of atomic absorption resulted in most papers on the subject being published in the spectrochemical, or at least the general analytical, journals. With its acceptance in all fields, articles now appear in the very specialized journals. It is therefore becoming more difficult to trace all references without the help of abstracts services or retrieval systems.

THE PRESENT STATUS OF ATOMIC ABSORPTION

When new physico-chemical analytical techniques are introduced, they appear to acquire, quite quickly, their own particular mythology. This may have the effect of perpetuating certain preconceived notions and, to some extent, hindering the search for the truth. A typical, early example in this field was that chemical interferences are fewer in atomic absorption than in emission. As both techniques depend upon the production of free atoms, they are, of course, the same. But a better subsequent knowledge of flame chemistry has aided understanding of interferences in flame emission, absorption and fluorescence. Other myths of a general nature which, hopefully, have now been refuted are that atomic absorption is a trace technique, that high temperature flames are dangerous and difficult to work with and that the quoted 'sensitivity' values tell you all you need to know about the performance of a particular instrument. Some sophisticated misconceptions undoubtedly still remain, though Alkemade[31] has started to explode some of these – in particular the often-quoted contention that atomic absorption is superior in sensitivity to atomic emission because the proportion of ground state atoms is higher.

Protagonists of the technique may sometimes give the impression that all other methods of elemental analysis are soon likely to be superseded. This also is a myth, for, while a chemist who invests several thousand dollars – or even pounds – will wish to extract as much information as possible from the equipment, he will also wish to use the technique that offers him most in terms of accuracy, cost and speed. Very often this is atomic absorption, but by no means always.

Comparison with other methods

When atomic absorption is compared with other methods of analysis – spectroscopic or otherwise – at least five capability factors must be taken into account. These are scope of application, ease of sample preparation, sensitivity and/or detection limit, reproducibility and accuracy.

As pointed out by Grant,[158] emission spectrography and spark source mass spectrometry would usually be the methods of choice for qualitative analysis involving metal ions. For quantitative analysis, atomic flame methods i.e. emission, absorption and fluorescence, come into their own where the sample already is, or is easily converted into, a liquid. Where a number of elements are to be determined together, direct reading emission or x-ray fluorescence multichannel spectrometers ('polychromators') offer the speediest, though much more expensive, solution. Such instruments may not provide optimum excitation conditions for all elements simultaneously however, and flame methods often offer better precision and accuracy. Sample heterogeneity limits accuracy in techniques where solid specimens are required, but, though this problem is overcome in solution, liquid containers and chemicals used in sample preparation are a cause of contamination. By contrast it is relatively easy to determine several elements sequentially by flame atomic methods, and this usually requires only one dissolution of the sample.

The concentration range where atomic absorption may be applied is from fractions of a part per million of many cations in solution up to tens of percent of metallic constituents in solid samples. Gravimetric methods with a precision of better than 1 part in 1000 still give higher accuracy in the 10–100% range. However, gravimetry is entirely non-specific and requires complete chemical separations. Titrimetry is usually designed to be specific, but with a precision of about 3 parts per 1000 is little better than the best atomic absorption results but is considerably less versatile though, of course, less expensive. Spectrophotometry usually covers a similar, perhaps slightly smaller, concentration range than atomic absorption. It has the advantage that its calibration curves shift less, and as the contents of the measurement cell are static, its precision and accuracy depend critically upon the quality of the instrument and the chemical separations and preparation of the sample – which are often time-consuming. Electrochemical methods compare with atomic absorption only when preliminary separations are not required.

Economic Factors. Economically, atomic absorption falls midway between the very inexpensive instrumentation such as colorimeters, balances, etc., and the very expensive, like automatic, direct reading spectrometers. But, as the sophistication of the instruments proceeds, relative costs are likely to rise. It is becoming customary to build in, or make provision for, such refinements as multilamp turrets, data-converting circuitry (so that results can be displayed or printed out direct in concentration), safety interlocks on the gas handling systems, automatic samplers, etc. These may well double the cost of a basic instrument. In the right circumstances this will soon pay for itself in doing the job of several other instruments, and in saving time and labour by avoiding a multiplicity of preparations of one sample for different elements. In every type of laboratory, versatility counts and many analysts have averred that the majority of metal determinations have been appreciably simpler and quicker since the introduction of the atomic absorption spectrometer.

Advantages. Lewis[237] summed up atomic absorption analysis by saying that it has a number of the advantages of an ideal technique: specificity; low limits of detection; many elements can be determined in one solution; there is rarely any 'lapsed time' requirement (as in colour development, drying of precipitates, etc.)

in sample preparation; and data output is in a directly readable form. To these we might well add the economic factor, versatility with regard both to types of sample and to concentration ranges and the fact that, though not an 'absolute' method, it can always be made entirely independent.

TERMS AND DEFINITIONS

During the growth period of a new analytical technique, existing terms may acquire a specialized meaning and certain new terms have to be coined. In order that all workers in atomic absorption may communicate with complete unanimity and understanding, a number of terms have been agreed by several international bodies. Most are based upon the definitions proposed by the Atomic Spectroscopy Group of the Society for Analytical Chemistry before the International Conference on Atomic Absorption Spectroscopy, Sheffield in 1969. Further discussions have resulted in the following definitions which, it is expected, will be adopted by the International Union of Pure and Applied Chemistry.

General terms

Atomic Absorption Spectroscopy. An analytical method for the determination of elements, based on the absorption of radiation by free atoms.

Atomic Fluorescence Spectroscopy. An analytical method for the determination of elements, based on the re-emission of absorbed radiation by free atoms.

Limit of Detection. The minimum concentration or amount of an element which can be detected with 95% certainty assuming a normal distribution of errors.

This is that quantity of the element which gives a reading equal to twice the standard deviation of a series of at least ten determinations at or near blank level.

Sensitivity. The sensitivity of an atomic absorption spectrometer is defined as: That concentration in solution of the analysis element which will produce a change, compared to pure solvent, of 0.0044 absorbance units (i.e. 1% absorption) in the optical transmission of the atomic vapour at the wavelength of the radiation used.

Noise Level. The noise level of an atomic absorption spectrometer is defined as: That concentration of the analysis element that would give a signal equal to one fiftieth of the sum of twenty measurements taken as follows:

> The output of an atomic absorption spectrometer operating on a blank solution is recorded for ten time periods each of ten times the time constant of the instrument. The maximum displacements that occur to both sides of the median line in each of the ten periods are measured. These are the twenty measurements referred to above.

The figure obtained approximates to the standard deviation of the noise expressed in terms of element concentration.

Terms concerned with the spectral radiation

Characteristic Absorbed Radiation. Radiation that is specifically absorbed by free atoms of the analysis element.

Resonance Radiation. Characteristic absorbed radiation that corresponds to the transfer of an electron from the ground state level to a higher energy level in the atom.

Radiation Generator. Equipment for producing characteristic absorbed radiation. The equipment normally consists of a lamp and power supply unit.

Hollow Cathode Lamp. A discharge lamp with a hollow cathode, usually cylindrical, used in atomic spectroscopy to provide characteristic radiation.

Boosted Output Lamp. A lamp in which a secondary discharge is used to increase the emission of characteristic absorbed radiation. This increase is normally both absolute and relative to other radiation from the lamp.

Electrodeless Discharge Tube (E.D.T.). A tube containing the element to be determined in a readily vaporized form and constructed so as to enable a discharge to be induced in the vapour. This discharge can be used as a source of characteristic radiation.

Terms concerned with the processing of the sample in an atomic absorption instrument

Atomic Vapour. A vapour that contains free atoms of the analysis element.

Atomization. The process that converts the analysis element, or its compounds, to an atomic vapour.

Atomizer. The device, usually a flame, used to produce and stabilize or maintain, a population of free atoms.

Sampling Unit. The part of an atomic absorption spectrometer (often consisting of a nebulizer, spray chamber, and burner) which accepts the sample solution and prepares it for atomization.

Nebulization. The process that converts a liquid to a mist.

Nebulizer. A device for the nebulization of a liquid.

Nebulization Efficiency. The ratio of the amount of sample reaching the atomizer to the total amount of sample entering the nebulizer.

Oxidant. The substance, usually a gas, used to oxidize the fuel in a flame.

Fuel. The substance, usually a gas, which is burnt to provide the atomizing flame.

Carrier Gas. The gas used to convey the sample mist to the atomizer.

Spray Chamber. The vessel wherein a mist is generated prior to transfer to an atomizer.

Flow Spoiler. A device, in a spray chamber, for removing large droplets from a mist.

Direct Injection Burner. A burner in which liquid is nebulized directly into the flame. The flame obtained with such a burner is normally turbulent.

Premix System. A sampling unit in which the fuel, oxidant gas and sample mist are mixed in a spray chamber before entering the flame. Flames obtained using this system are normally laminar.

Semi-premix System. A sampling unit in which either fuel or oxidant gas is added to the sample mist after the spray chamber.

Long Tube Device. A device in which an atomizing flame is directed into a tube lying along the optical axis of an atomic absorption spectrometer.

Long Path Burner. A burner constructed to produce a flame which is extended in one direction at right angles to the direction of movement of the flame gases.

Multislot Burner. A burner with a head containing several parallel slots.

Separated Flame. A flame in which the diffusion combustion zone is so separated from the primary combustion zone as to enable the two zones to be observed independently. Separation may be effected either mechanically or by an inert gas shield.

Observation Height. The vertical distance between the optical axis of the monochromator and the top of the burner.

Burner Angle. The acute angle between the plane of the flame produced by a long path burner and the optical axis of the monochromator.

Detectors

Resonance Radiation Detector. A selective detector in which atoms in an atomic vapour are excited by radiation from an external source, and the intensity of the resulting fluorescence radiation is measured.

Terms concerned with the analytical technique

Interference. A general term for an effect which modifies the instrumental response to a particular concentration of the analysis element.

Depression. An interference that causes a decreased instrument response.

Enhancement. An interference that causes an increased instrument response.

Matrix Effect. An interference caused by differences between the sample and a standard containing only the analysis element and, where appropriate, a solvent.

Radiation Scattering. An interference effect caused by scattering of radiation from drops or particles associated with the atomic vapour.

Spectroscopic Buffer. A substance which is part of, or is added to, a sample and which reduces interference effects.

Ionization Buffer. A spectroscopic buffer used to minimize or stabilize the ionization of free atoms of the analysis element.

Releasing Agent. A spectroscopic buffer used to reduce interferences attributable to the formation of involatile compounds in the atomizer.

Comments on the definitions

The relationship between the terms 'sensitivity' and 'detection limit' is discussed in Chapter 4 (under Trace Analysis). In this sense it is restricted to describing the capability of an instrument in respect of a specified element. The word sensitivity may also be used in a general non-quantitative sense when describing a method or even an instrument.

The words 'atomizer' and 'atomization' are used in their entirely literal sense: a thing or a process which turns a substance into atoms. These terms should now never be used in connexion with the scent-spray-like device used for producing a mist from a liquid.

The term 'support gas' is still used by different writers to mean either the gas which supports the aerosol, or the gas which supports combustion. To avoid confusion 'carrier gas' and 'oxidant' are proposed instead.

Chapter 2

Basic Principles

EMISSION, ABSORPTION AND FLUORESCENCE SPECTRA

The analyst about to use atomic absorption or fluorescence spectroscopy will find it desirable to know something about the way in which atoms react with various forms of energy – particularly light and heat.

Under the conditions in which atomic spectroscopy, emission, absorption and fluorescence, is practised the energy input into the atom population includes thermal, electromagnetic, chemical and even electrical forms of energy. These are converted to light energy by various atomic and electronic processes before being measured. The light energy is manifest in the form of a spectrum, which consists of radiation of a number of discrete wavelengths.

First Balmer and then Rydberg described the spectrum of hydrogen in terms of exact mathematical formulae, which could be extended to other univalent atoms such as the alkali metals. The Rydberg equation, which expresses the wavenumber of a given spectrum line as the difference between two *terms,* one term being constant for a given series of spectrum lines, led to the old quantum theory of atomic spectra in which discrete or quantized energy levels are postulated for the Coulomb forces between the valency electrons and the positive nucleus of the atom. Transition between two such quantized states corresponds to the absorption or emission of energy in the form of electromagnetic radiation, the frequency ν of which is determined by Bohr's condition:

$$\Delta E = E_1 - E_2 = h\nu$$

where E_1 and E_2 are the energies in the initial and final states respectively, and h is Planck's constant.

Both the quantum theory and the more recent wave mechanics theory of Schrödinger have been developed further to correlate atomic structure with atomic spectra, but such detailed knowledge is not necessary for following the processes on which the present subject is based.

An atom is said to be in the ground state when its electrons are at their lowest energy levels. When energy is transferred to a population of such atoms by means of thermal or electrical excitation, as in the various forms of emission spectroscopy, the transfer takes place by means of collision processes. The amount of energy transferred may vary considerably from atom to atom, resulting in a number of different excitation states throughout the population. The subsequent emission of these different amounts of energy involves not only many higher energy levels, but also low energy levels of other than the ground state. These result in radiation of a number of different frequencies, and hence the *emission* spectrum of any given element may be highly complex.

Theoretically, the reverse process is also possible. That is, if light of any of these given frequencies is passed through a vapour containing the atoms, it will be absorbed in performing the process of excitation. However, the proportion of excited to ground state atoms in a population at a given temperature can be considered with the aid of the well known Boltzmann relation

$$\frac{N_m}{N_n} = \frac{G_m}{G_n} \exp\left[-\frac{(E_m - E_n)}{kT}\right]$$

where N is the number of atoms in a state n or m, G is the statistical weight for a particular state and k is the Boltzmann constant. Walsh[415] has calculated the ratio N_m/N_n for a number of common atoms over a range of temperature, see Table 1, for the case where m refers to the first excited state and n is the ground

TABLE 1. Values of N_m/N_n for typical elemental resonance lines

Line, nm	G_m/G_n	N_m/N_n			
		2000 K	3000 K	4000 K	5000 K
Cs 852.1	2	4.44×10^{-4}	7.24×10^{-3}	2.98×10^{-2}	6.82×10^{-2}
Na 589.1	2	9.86×10^{-6}	5.88×10^{-4}	4.44×10^{-3}	1.51×10^{-2}
Ca 422.7	3	1.21×10^{-7}	3.69×10^{-5}	6.03×10^{-4}	3.33×10^{-3}
Zn 213.9	3	7.29×10^{-15}	5.58×10^{-10}	1.48×10^{-7}	4.32×10^{-6}

state. The very low proportion of atoms in the first excited state, even at temperatures of 3000 K indicates that absorption of radiation, other than that originating from a transition involving the ground state, would be very small.

In the sun's spectrum many such absorptions show up as Fraunhofer lines, but in laboratory experiments, and certainly on the analytical scale, only absorptions involving the ground state are normally observed. Absorptions involving the ground state are for this reason known as *resonance* lines.

The absorption spectra of most elements, therefore, when produced under laboratory conditions, are extremely simple. This accounts for one of the main advantages of atomic absorption spectra as a means for analysis – there is very little possibility of coincidence of lines and therefore very little spectral interference.

Atoms excited by absorption of resonance radiation also re-emit the absorbed energy. This process, by analogy with a similar process known in molecular spectroscopy, is called atomic fluorescence. The re-emitted energy may however be of the same wavelength as the absorbed energy, or it may be of longer wavelength, indicating that an intermediate state is also involved, with the consequent partial loss of energy in some other form.

Direct re-emission, (a) in Fig. 2, is known as resonance fluorescence, and two forms of indirect re-emission, involving fluorescence to an intermediate state (b), and excitation to a state higher than the first excited state (c), are referred to as

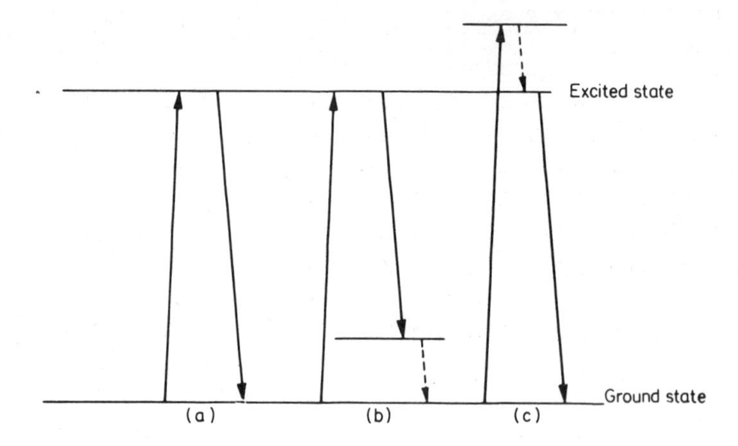

FIG. 2. (a) Resonance fluorescence. (b) Direct line fluorescence. (c) Stepwise fluorescence.

'direct line' and 'stepwise' fluorescence respectively. 'Sensitized fluorescence' in which an atom is first excited by transfer of energy from an atom of another element, itself excited by absorption of radiation, is also known.

The fluorescence spectrum of a given element, when excitation is by absorption of resonance radiation, is even simpler than the absorption spectrum, and consists of only a very small number of lines.

As the means used for producing atomic vapours always result in the production of heat, all three processes occur, to some extent, simultaneously though they can be observed and measured separately by appropriate choice of the experimental conditions.

The very high proportion of ground state to excited state atoms when these are in equilibrium, as shown in Table 1, suggests two particular advantages of atomic absorption and fluorescence measurements over emission. Because absorption and fluorescence are direct measurements of the number of ground state atoms, these would be expected to yield a better sensitivity than emission. This is only one factor of several which must be taken into account when absorption and emission sensitivities are compared, and it is rarely the most important. However, as the temperature coefficient of the number of excited atoms is markedly greater than that of the ground state atoms, absorption and fluorescence measurements would be expected to be less dependent upon short-term flame temperature variations than emission. This is indeed reflected in their generally better signal-to-noise ratios.

Alkemade[31] has shown that for a given analysis line, the absorption sensitivity is better than the emission only if the spectral radiance of the lamp source exceeds that of a black body at the temperature of the flame. Since the sharp line sources used in atomic absorption are invariably excited by non-thermal means, their spectral radiance exceeds that of a flame atomizer by many orders of magnitude.

The effective radiation temperature for resonance lines produced in gas discharge tubes, for example, is about 10000 K.

A factor which also has practical implications is the wavelength of a resonance line, which is inversely proportional to its energy ($\Delta E = h\nu = hc/\lambda$). A result of the Boltzmann relationship (page 9) is therefore that the ratio N_m/N_n decreases exponentially with increasing resonance line frequency. Thus it is expected, and found, that elements whose resonance lines occur at the higher wavelengths are more sensitive in emission than those whose resonance lines are at low wavelengths. Lithium (670.7 nm) and sodium (589.0 nm), for example, are more sensitive in emission than copper (324.8 nm) or magnesium (285.2 nm), while zinc (213.9 nm) is very insensitive indeed in emission.

The converse is not generally true for absorption, for although the absolute amount of *energy* absorbed is directly frequency dependent, under experimental conditions one only measures the *proportion* of the incident energy which is absorbed. The frequency term thus cancels out leaving an expression of the form given below.

RESONANCE RADIATION

In analytical atomic absorption and atomic fluorescence spectroscopy, one is dealing largely with the absorption and emission of resonance radiation which we define as characteristic radiation of an element that corresponds to the transfer of an electron from the ground state to a higher energy level. A comprehensive treatise on the interactions between atoms and resonance radiation was compiled by Mitchell and Zemansky.[271]

The analytical validity of making absorption measurements depends on the relationship between absorption and the concentration or partial pressure of the absorbing atoms. This relationship is, according to classical dispersion theory

$$\int K_\nu \, d\nu = \frac{\pi e^2}{mC} N_\nu f$$

where K_ν is the absorption coefficient at frequency ν, m = electronic mass, e = electronic charge, and N = number of atoms per cm^3 capable of absorbing energy in the range ν to $\nu + d\nu$. The f-value, now called the oscillator strength, is, for the present purpose, the effective number of free electron oscillators per atom of the element in question whose absorption effect is equivalent to that produced by the incident radiation. From the data of Table 1 it is clear that the value N – the total number of atoms of the element present – can with negligible error be substituted for N_ν. (This relationship holds unless the ground state is a multiplet state, in which case the expression is slightly more complex.) Nevertheless, the energy absorption is virtually a direct function of the number of atoms in the absorbing path.

The natural width of an atomic spectral line is of the order of 10^{-5} nm (10^{-4} Å), but this is broadened by Doppler, electric field and pressure effects. Doppler widths vary with the element, the wavelength of the line and the temperature. The sodium 589 nm line for instance, has a Doppler width of 0·0048 nm at 3,000 K, but if the line is produced in a flame, the pressure effect may broaden it to about 0·008 nm.

As Walsh pointed out when first seriously examining the potentialities of atomic absorption as an analytical technique,[415] the practical implication of this is that, in order to measure the absorption coefficient of a given line in a white continuum, a spectrograph with a resolution of something like 500 000 would be necessary. This is hardly practicable for the reason that the energy passed by the required small spectral slit-width would be too low to be measured by standard photoelectric methods although limited use has been made of photographic

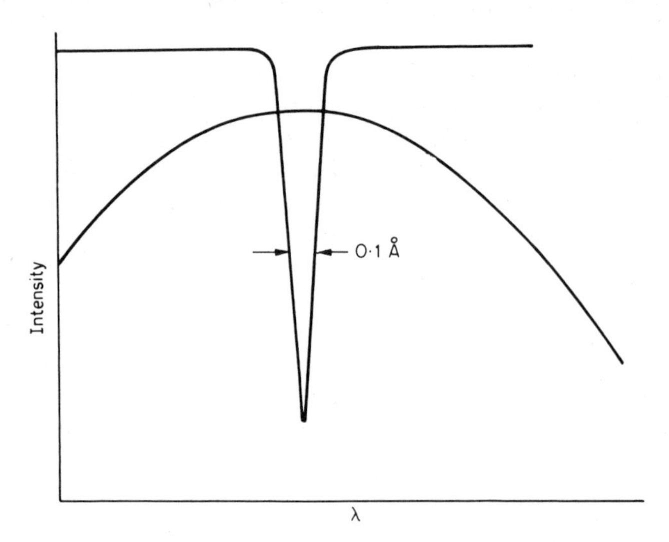

FIG. 3. Absorption line and monochromator bandpass.

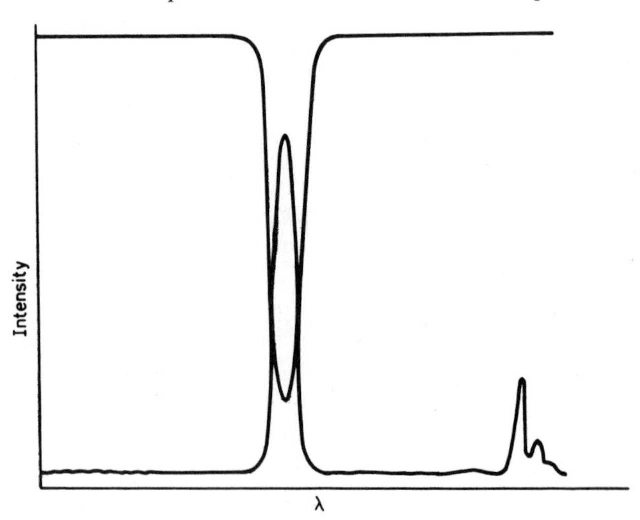

FIG. 4. Absorption line and fine emission line.

recording. Measurements made over wider band widths than this result in absorption values smaller than the true values, though within certain limits and under controlled conditions these can be put to some analytical use.

However, the absorption coefficient can be measured over an area close to the centre of the absorption line with a sharp-line source which emits lines of smaller half-width than the absorption line itself. If the shape of the absorption line is determined solely by Doppler broadening, then a linear relationship can still be shown to exist between the concentration and the energy absorbed. With a sharp-line source, it becomes unnecessary to use a monochromator bandwidth of the same order as the half-width of the absorption line. A spectrometer is needed simply to isolate the line to be measured from the other lines emitted by the source.

The position is clarified in Figs. 3 and 4. In Fig. 3 the absorption line profile is superimposed on a broader profile, which may represent either a non-sharp emission line, or even, for the purpose of the present argument, the bandpass of a laboratory spectrometer of normal resolution. The proportion of energy actually absorbed is such a small fraction of the total that, at best, the measurement is highly insensitive. When the absorption is superimposed on the sharp emission line, energy in the latter is absorbed over the full width of the emission line and the measured absorption then approaches the theoretical maximum (Fig. 4).

In order to measure the amount of absorption by atoms of a given element, it is necessary to devise an experimental arrangement such as that shown in plan in Fig. 5. A light-measuring device placed at position A is enabled to indicate by

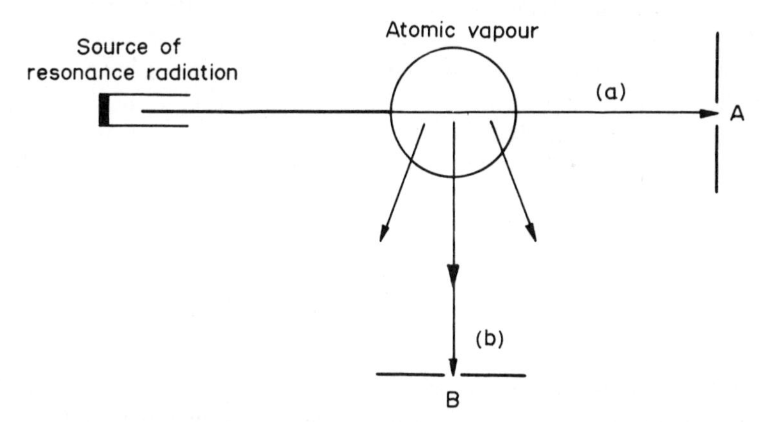

FIG. 5. (a) Primary beam attenuated by absorption. (b) Fluorescence emission.

difference the amount of energy absorbed by the atomic vapour, while one placed at any other position, e.g. at B, will measure a proportion of the light which is re-emitted as fluorescence.

It is important that the primary source should be a narrow line emitter for absorption purposes, the 'wings' of the spectral profile of a source of wide bandwidth being unabsorbable light entering the detection system. But the same constraint does not apply to the case of fluorescence, as only the fluorescence radiation itself (plus some scatter) enters the detector. In a rigorous treatment of the subject the fluorescence intensity is shown to be a function of the fluorescence efficiency, and therefore of the primary source line width, but the effect of increased line width on fluorescence measurements is minor compared with the effect of unabsorbed radiation on absorption measurements.

PRODUCTION OF FREE ATOMS

All the preceding matter referred to the resonance spectra of free, un-ionized atoms. It is on the production of uncombined and un-ionized atoms in an atomic vapour that the success of an atomic absorption or fluorescence analytical procedure depends.

Consider the events that take place when fine droplets of a solution are aspirated into a plasma at a temperature of 2000–3000 K. First, the solvent evaporates leaving small solid particles (referred to by some as 'clotlets'). These particles melt and vaporize. The vapour consists of a mixture of compounds which tend to decompose into individual atoms. Free atoms obviously exist, however transitorily, under these conditions, otherwise there would be no evidence of atomic spectra. The individual atoms absorb energy by collision and become excited or ionized. Taking potassium chloride as an example, the two last stages of this process may be written:

$$KCl \rightleftharpoons K + Cl$$
$$K \rightleftharpoons K^+ + e^-$$

and it will be seen that, while the first reaction provides the free atoms required to ensure sensitivity of the method, the second reaction tends to remove them.

Both the molecular dissociation and ionization processes have been investigated thermodynamically to some extent, but it would be difficult and perhaps impossible to give a comprehensive treatment for all the elements likely to be encountered in atomic absorption. Degrees of atomization differ markedly from element to element in a given plasma. It was pointed out by de Galan and Samaey[148] that this factor more than any of the others (oscillator strength, line broadening, etc.) influences the analytical sensitivity. The following simple and partially qualitative approach will at least give some idea of the significance of the atomization and ionization processes, and enable many of the observed effects to be explained.

Diatomic dissociation and ionization are similar processes, and Gaydon[152] using the Saha equation[342] has calculated the degree of dissociation of atoms into ions and electrons for various elements, temperatures and partial pressures. Some of these are plotted for potassium giving the curves I, II and III in Fig. 6, and for calcium, curve I in Fig. 7.

For the elements calcium and potassium, a partial pressure of free atoms in the flame of 10^{-6} atmosphere corresponds very approximately to the amount of these elements passing into the flame when a conventional atomic absorption spectrometer is used (the sample flow rate being 3 ml per minute, and the gas and air flows 1.5 l and 6 l per minute respectively) to analyse a solution containing 100 p.p.m. A more practical value would be 10^{-8} atmosphere, corresponding to 1 p.p.m. of calcium or potassium in the sample solution. The estimated curve for this value for potassium is curve IV, Fig. 6.

The degrees of dissociation of simple compounds, such as potassium chloride,

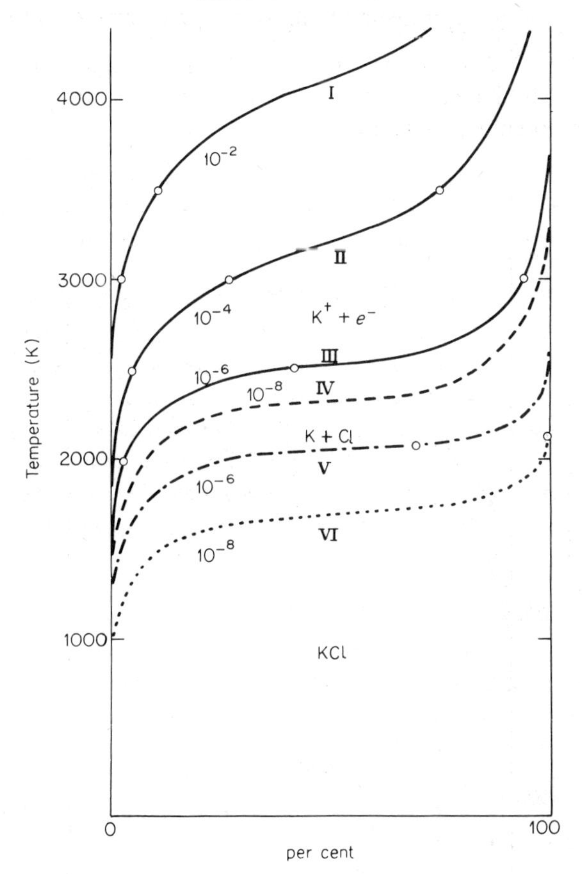

FIG. 6. Atomization and ionization curves for potassium.

follow a similar pattern. Mavrodineanu[258] and Rubeška and Moldan[4] have quoted the dissociation constants of some alkali halides at 2000 K. From their values for potassium chloride [(pK × pCl)/pKCl = 2.5×10^{-6} at 2100 K and 7.9×10^{-5} at 2400 K], points through which the dissociation/temperature curve passes for each of the two partial pressures 10^{-6} and 10^{-8} are plotted on Fig. 6, and probable curves V and VI drawn in.

The true position of the curves depends on very many factors, including the nature of the flame gases, the presence of other ions both positive and negative in the sample solution and the formation of free radicals from the breakdown of the solvent. However, from the pair of curves relating to a partial pressure of 10^{-8}, the proportion of free potassium atoms which would be present under interference-free conditions can be deduced for any temperature. This is a maximum where the curves show the maximum horizontal separation, i.e. at about 2000 K. This is in good agreement with the observed fact that the maximum sensitivity for potassium in both emission and absorption is obtained with an air propane flame giving approximately this temperature.

It appears that for much higher concentrations of potassium (e.g. 10^{-6} atmosphere partial pressure or 100 p.p.m.) the best sensitivities would be given at higher flame temperatures. Use is unlikely to be made of this observation because the measurement of such high concentrations, both in emission and absorption, poses certain other instrumental problems.

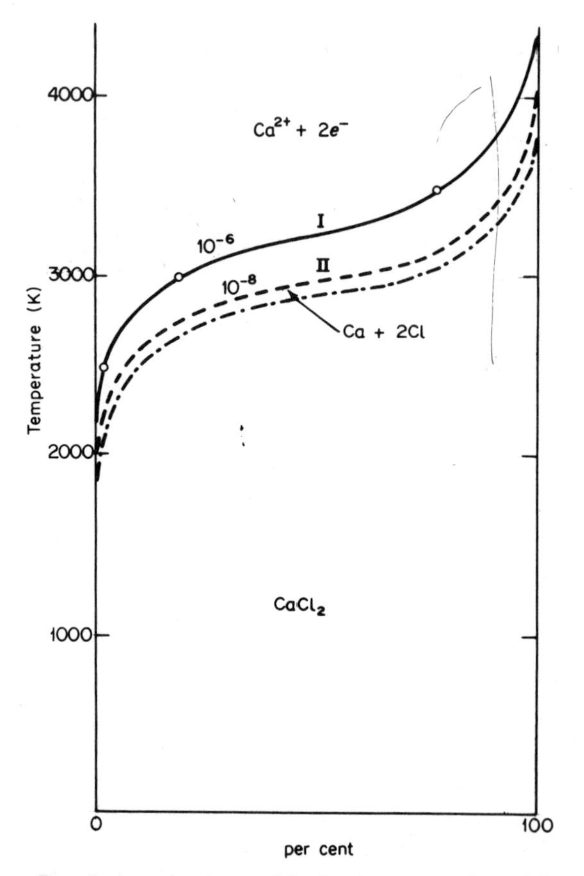

FIG. 7. Atomization and ionization curves for calcium.

For calcium only the 10^{-6} atmosphere partial pressure curve (Fig. 7 curve I) has been fully drawn in, as higher concentrations are again of little practical interest. The curve for 10^{-8} atmosphere is probably near curve II.

A molecular dissociation curve for calcium cannot be placed with any confidence, though the degree of atomization for calcium in the air acetylene flame, as calculated by de Galan and Samaey[148] suggests it could be the third curve. This is supported by the higher sensitivity for calcium in higher temperature flames and also explains the critical dependence of calcium sensitivity on flame parameters. The presence of other ions and flame radicals affects the dissociation constant and thus influences the vertical position of this curve. They also therefore influence the horizontal separation of the atomization and ionization curves, which is a measure of the number of free atoms. The practical significance of these dissociation curves will be more apparent when the flames themselves and the so-called interference effects between the species present or produced in them are discussed in Chapter 4 (p. 86).

Chapter 3

Instrumental Requirements

THE BASIC SYSTEM

In order to be able to make measurements in atomic absorption it is necessary to devise an experimental assembly which will convert the material under examination as efficiently as possible to a population of ground state atoms, and then pass resonance radiation of the element to be measured through that population. Ideally, the light-measuring device should 'see' only the wavelength which is being absorbed, as the presence of other radiation will lower the proportion of absorbed radiation, and thus decrease the sensitivity of the measurement.

The basic atomic absorption spectrometer is shown in Fig. 8. Light from the

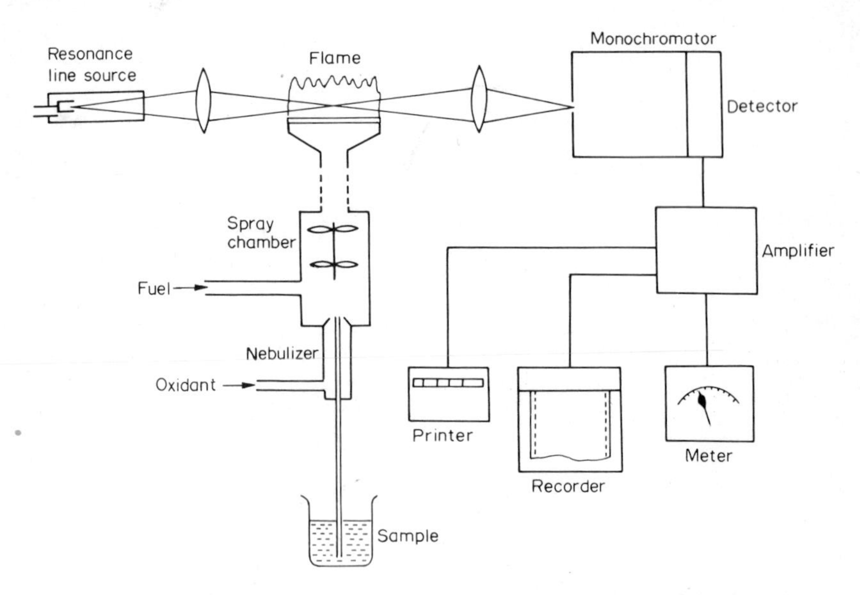

FIG. 8. Practical system for atomic absorption spectrometer.

source lamp generating a sharp line spectrum characteristic of the desired element passes through the flame into which the sample solution is sprayed as a fine mist. The region of the spectrum in the immediate neighbourhood of the resonance line to be measured is selected by the monochromator. The isolated resonance line falls on to the detector, a photomultiplier, the output of which is amplified and drives a readout device e.g. a meter, strip chart recorder or a digital display unit or printer. The intensity of the resonance line is measured with and without the sample passing into the flame. The difference between these readings, as will be discussed in detail on page 77 is a measure of the absorption and therefore of the amount of the element being determined.

In order to avoid measuring the emission from the excited atoms in the flame at the same wavelength, the source lamp intensity is modulated, usually at 50 Hz or sometimes at 400 Hz, and the amplifier is tuned to the same frequency. Consequently, the continuous component of the radiation signal, originating from the flame, is not measured.

The same or similar components may, in principle, be used to make measurements in flame emission and atomic fluorescence spectrometry. For emission, the source lamp is removed, and the flame radiation modulated by a reed or rotating sector wheel at the amplifier frequency. In fluorescence, the source radiation is passed through the flame at an angle, usually a right angle, to the optical axis of the monochromator. The source is still modulated in order to avoid measuring thermally excited emission from the flame.

Instrumentation for atomic spectroscopy – emission, absorption and fluorescence – thus has two parts with essentially different functions, the means whereby the population of ground state atoms is produced from the sample, and the optical system which includes the resonance source and the spectrometer.

THE PRODUCTION OF ATOMIC VAPOURS

Combustion flames in premix systems

In spite of much development in electrothermal atomization, the most convenient, stable and economic source of atomic vapours remains the combustion flame.

Fuel/oxidant mixtures are now commonly used which are both safe to handle and which produce a range of temperatures from about 2000 to 3000 K. As we have seen, not only is the flame temperature an important parameter in the production of free atoms from a given element but also the chemical effects of radicals and other substances present in the flame. One gas mixture producing a given temperature can have quite different analytical properties from another mixture either of the same or of different gases producing the same temperature.

Fuel gases include propane, hydrogen and acetylene, and oxidants include air and nitrous oxide. Pure oxygen is rarely used as an oxidant at the present time as this provides a mixture whose burning velocity is high and difficult to control, but oxygen is often mixed with an inert gas such as nitrogen or argon.

Fuel/oxidant mixtures may be combined from most of the above in ratios which may be stoichiometric, lean or rich. Lean and rich mixtures contain less than, and more than the stoichiometric quantity of fuel gas respectively. The characteristics of some frequently used flames are summarized in Table 2. The flow rates given may not be those that would be shown on flow meters calibrated for air. These are not directly proportional to the quantities of reactants in the stoichiometric flame, as reading is dependent on the density of the gas, and diffusion of oxygen from the atmosphere affects the flow ratio at which stoichiometry occurs.

TABLE 2. Characteristics of some pre-mixed gas combustion flames

	Flow rates l/min Fuel	Oxidant	Approx. temp. K	Expansion factor	Flame speed cm/sec
Air propane					
lean	0.3	8			
stoichiometric	0.3–0.45	8	2200		45*‡
rich	0.45	8			
Air acetylene					
lean	1.2	8			
stoichiometric	1.2–1.5	8	2450	1.03	160*‡
luminous	1.5–1.7	8			
rich	1.7–2.2	8	2300		
N_2O acetylene					
lean	3.5	10			
stoichiometric	3.5–4.5	10	3200	1.64	285†
rich	4.5	10			
Air hydrogen stoichiometric	6	8	2300	0.9	320*‡
N_2O hydrogen stoichiometric	10	10	2900	1.00	380‡
N_2O propane stoichiometric	4	10	2900		250

* Gaydon and Wolfard[152] † Aldous *et al.* [465] ‡ Willis [469]

Undoubtedly the most widely used of these fuel/oxidant mixtures is air acetylene as it enables about 30 of the common metals to be determined. The sensitivity of some elements in the hydrogen air flame is not much inferior to their sensitivity in air acetylene (i.e., their degrees of atomization are comparable) but the presence of other substances in the sample solution may cause the formation of stable compounds, and hence interferences are said to be worse. In particular, elements which form stable monoxides give lower sensitivity in hydrogen-based flames.

Air propane mixtures were used extensively in the earlier days of atomic absorption analysis for those elements which were found to be easily atomized. These included the alkali metals, cadmium, copper, lead, silver and zinc. The air propane flame does indeed appear to give a better sensitivity for these elements, though other factors contribute to make the detection limit at least no better, and perhaps somewhat worse, than when air acetylene is used.

Air acetylene mixtures are used successfully for most elements which do not form highly refractory oxides. Calcium, chromium, iron, cobalt, nickel, magnesium, molybdenum, strontium and the noble metals are among the elements normally determined with this flame. The refractory oxide metals may give better sensitivity in a fuel-rich flame – though this on the average is about 150° C cooler than a stoichiometric flame. Elements which give very low sensitivity in air acetylene flames (and for which this flame analytically is of little use) are those whose dissociation energies for the M–O bond are greater than about 5 eV. Examples are: Al–O, 5.98; Ta–O, 8.4; Ti–O, 6.9; Zr–O, 7.8 eV.

The use of nitrous oxide as the oxidant gas instead of air was suggested by Willis.[445] Its performance under analytical conditions was reported by Amos and Willis.[36] This gas might be expected to provide properties halfway between air and pure oxygen. Mixed with acetylene however, nitrous oxide produces a higher temperature than the equivalent mixture of oxygen and nitrogen, due mainly to the exothermic nature of its decomposition reactions. Temperatures of about 3000 K are produced, not much less than with oxy-acetylene, but the flame propagation is slower. The flame is thus safer to handle than oxy-acetylene.

Nitrous oxide has also been used as oxidant for hydrogen and propane. The nitrous oxide hydrogen flame has little to commend it for atomic absorption purposes because of the surprisingly low sensitivities which it gives. Chester, Dagnall and Taylor[86] found that the nitrous oxide hydrogen flame is highly oxidizing in nature as compared with the nitrous oxide acetylene flame. While it is useful for some elements in overcoming interference effects, very low sensitivities are given for elements which form refractory oxides. Nitrous oxide propane or butane have been shown by Butler and Fulton[80] to have certain of the interference-reducing properties of the nitrous oxide acetylene flame, while retaining the ease of handling of air acetylene, though again, sensitivities in general are disappointing.

There seems little doubt, then, that the two most useful flame mixtures will continue to be air acetylene and nitrous oxide acetylene.

A recent study by de Galan and Samaey[148] of the performance of flames based on air or nitrous oxide with hydrogen or acetylene has led to some valuable observations. Best sensitivity nearly always occurs just above the inner cone of the laminar flame. This is usually 0.3–0.5 cm above the burner slot. The air acetylene and nitrous oxide acetylene flames are in thermal equilibrium at this point, whereas the air hydrogen flame attains thermal equilibrium only at about 15 cm above the slot. Better sensitivities might thus be observed at a point in the hydrogen flames which unfortunately is adversely affected by physical instability.

Two main factors arising from the conditions in the flame itself affect the degree of atomization and ultimate sensitivity of a particular metal.

Gas flow rates and thermal expansion of the flame gases after combustion both contribute to the total dilution of the absorbing species in the flame and to the length of time (usually no more than 10^{-+} second) which an absorbing atom actually remains in the radiation beam, and the best sensitivity at a given temperature is given by a flame having the smallest combustion gas/sample volume ratio. At the lower end of the flame temperature scale this is the air propane flame, which explains why, for elements completely atomized at 2200 K, this flame gives better sensitivities than air acetylene. For hydrogen-based flames the volume of fuel is comparatively greater than for hydrocarbon-based flames. This is part of the explanation of the poorer performance of nitrous oxide hydrogen as compared with acetylene nitrous oxide.

Isothermal degrees of atomization have been shown by Chester et al.[86] to be proportional to the concentration of carbon and carbon-containing species and inversely proportional to the concentration of atomic and molecular oxygen in the flame. The presence of free carbon thus exerts the major effect on the reducing properties of the flame. In a given hydrocarbon flame, as the fuel/

oxidant ratio is increased, the concentration of free carbon or small carbon-containing radicals increases up to the point of formation of soot (massive carbon agglomerates which effectively remove the carbon activity). At this transition point occurs the critical carbon/oxygen ratio. For certain elements, the sensitivity is observed to increase to a maximum at the critical C/O ratio. The presence of organic solvents increases the critical C/O ratio and thus for metals in solution in organic solvents higher sensitivities are usually found than for aqueous solutions. The critical C/O ratio can be shown to be twice as high for acetylene as for propane, and of course for hydrogen combustion mixtures the effect of carbon activity is absent.

The success of nitrous oxide acetylene as an atomizing gas in atomic absorption may therefore be ascribed to the high temperature produced by nitrous oxide as oxidant and the high critical C/O ratio contributed by acetylene.

Nebulizer – burner systems

The prime purpose of these systems is the conversion of the sample solution into the atomic vapour where the absorption measurement is made. In a sense therefore, this is the heart of the instrument, for upon its correct function and efficiency the sensitivity of the analysis depends directly.

The processes include nebulization (i.e., the conversion of the liquid sample to a mist or aerosol), the selection of mist droplets of the correct size distribution,

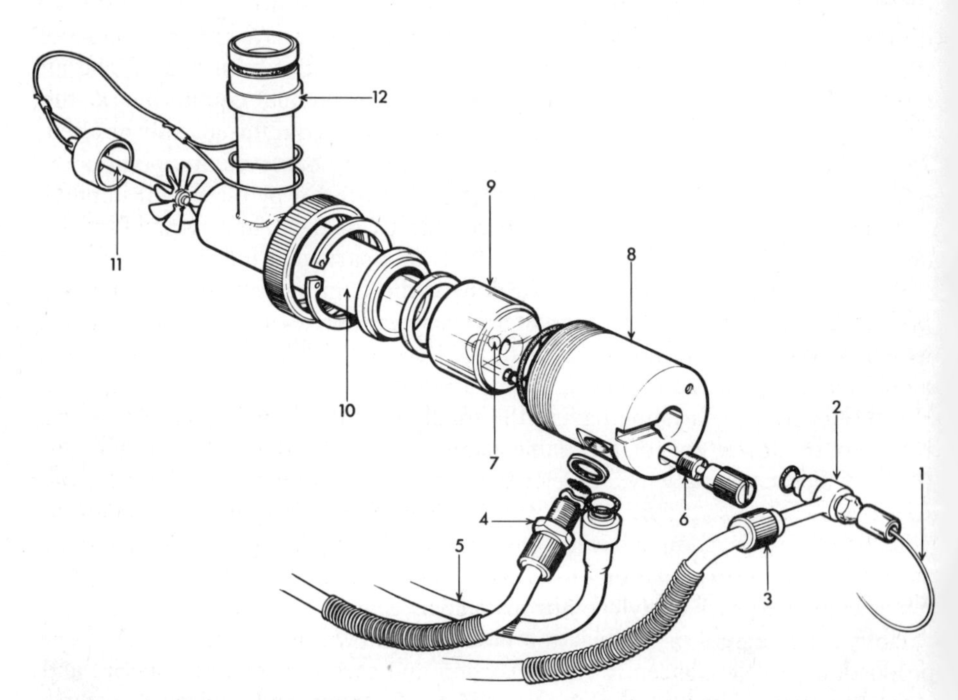

FIG. 9. Sampling unit (from Unicam SP1900). 1. Sample uptake capillary; 2. Nebulizer; 3. Oxidant inlet; 4. Fuel inlet; 5. Drain; 6. Bead holder; 7. Impact bead; 8. Steel protective body; 9. Inert insert; 10. Spray chamber; 11. Baffle and safety bung; 12. Burner stem.

the mixture of the selected mist with the flame gases and introduction to the burner. These are the processes which occur in pre-mix spray chamber type burner systems. The merits of such systems as opposed to direct injection types have been well established during the past ten years or more, and spray chambers are now almost universally employed in atomic absorption spectrometers, in spite of their comparatively low nebulization efficiency.

A typical system is shown in exploded view in Fig. 9. The sample is drawn into the nebulizer by the low pressure created around the end of the capillary by the flow of the carrier or oxidant gas. The resulting droplets are ejected with the carrier gas into the spray chamber. The design of the chamber is such that droplets with a diameter greater than about 5 μm fall out onto the sides of the chamber and flow to waste. The fuel gas is introduced into the chamber, and also auxiliary carrier gas or oxidant, so that an intimate mixture of sample mist, fuel and oxidant leave the spray chamber and enter the burner. A laminar flow burner sustains a highly stable flame, and the proportion of sample mist normally introduced has little or no influence over the flame characteristics. Two exceptions to this last statement are when organic solvents (which alter the carbon/oxygen ratio) are employed, and when the volume ratio of liquid sample and combustion gases exceeds a critical ratio of about 1:5000. Despite the complexity of this system, the response time is fast, and only about one second elapses between the introduction of a sample solution and the signal reaching the readout system. A steady reading, corresponding to dynamic equilibrium in the flame, should be reached in 7–10 seconds.

The particular advantage of the spray chamber system is that homogeneous combustion, ensured by premixing the gases, makes for a nonturbulent or laminar flame. Good stability results together with well defined temperature zones, so that the region of the flame (observation height) where best sensitivity is produced can be selected.

In direct injection burners, the nebulization process takes place in the burner itself. The oxidant, which draws in the sample, is also mixed with the fuel gas in the flame, resulting in non-homogeneous combustion and a high degree of turbulence. Such burners are not well adapted to give the long narrow flames which produce highest sensitivities in atomic absorption.

Pneumatic nebulizer

A nebulizer unit is shown in section in Fig. 10. For a given gas pressure the efficiency and the droplet size distribution given by this device depend almost entirely on the capillary diameter and the relative positions of the end of the capillary and the nose-piece. Best performance may thus be attained by careful adjustment of the position of the capillary usually by the screw insert which holds it.

Although nebulizer units are constructed robustly to withstand mechanical deformation, they may be damaged by acid attack on the surfaces exposed to the sample solutions. The incidence of acid solutions at high speed with a high proportion of entrained air are well known to be very highly corrosive. Stainless steel will withstand mild acid attack under these conditions but will be rapidly dissolved away by solutions containing 1% ferric ions in hydrochloric–nitric acid. The best nebulizers are therefore fabricated from much more inert material.

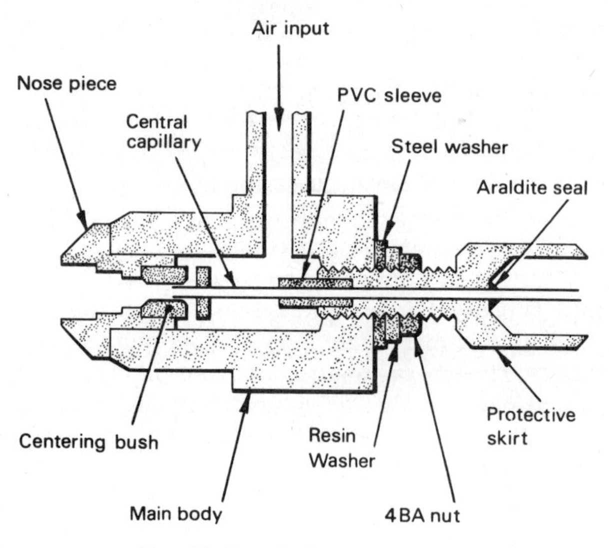

FIG. 10. Detail of nebulizer.

Platinum–iridium alloy is often used for the central capillary and tantalum or platinum for the nose-piece and annulus.

Nebulizer units are designed to operate at maximum efficiency (i.e. give the largest proportion of fine droplets) for a given oxidant flow rate. If the flow rate is either reduced or increased the efficiency is likely to be impaired, and at higher flow rates the analytical sensitivity will not increase in proportion, but may well decrease. Design parameters include the capillary dimensions and nose-piece geometry.

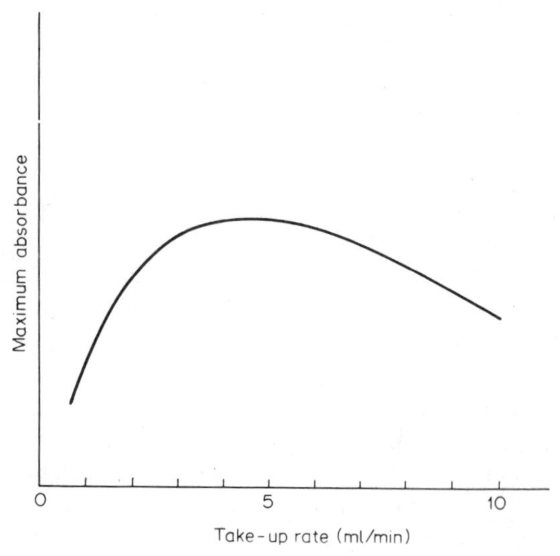

FIG. 11. Characteristics of pneumatic nebulizer.

There appears also to be an optimum solution take-up rate for this basic form of nebulizer. If, for a series of nebulizers designed to give maximum efficiency at various take-up rates, the best absorbance is plotted against the take-up rate, a curve envelope of the form shown in Fig. 11 is obtained. The maximum occurs between flow rates of 3 and 6 ml/minute and corresponds to a nebulization efficiency of 10%. Higher absorbance can therefore never be obtained by further adjustment.

These particular values apply to dilute aqueous solutions, but the nebulization rate is also affected by variations in sample viscosity and surface tension and by the equilibrium temperature of the spray inside the spray chamber.

The actual sizes of droplets produced by a nebulizer vary very widely, and the diameters range from <5 μm to 25 μm or more. The purpose of the spray chamber is to limit the size of droplets reaching the burner to just those which can be vaporized and atomized in the flame (i.e. of the order of 10 μm or less) and thus contribute to analytical sensitivity. If a spray chamber prevents vaporizable droplets from entering the flame analytical sensitivity will be decreased, while if it allows unvaporizable droplets to reach the flame, the flame noise signal will be increased, and the flame temperature is reduced. The maximum useful population of droplet sizes constitutes about 10% of the total mass of sample nebulized, and this is why this figure also represents the maximum attainable efficiency already quoted.

To improve the nebulization efficiency various ways and means of altering the normal droplet size distribution have been employed. Perhaps the best known of these is the application of heat, either to the sample and gases before they enter the spray chamber, or in the spray chamber itself. Heating of the sample produces a minimal effect until the temperature approaches the boiling point. The reduced pressure applied to the sample solution in the nebulizer then results in pre-boiling and erratic nebulization. It has been proved[320] that the heating of the oxidant gas can result in a useful increase in sensitivity, but unless the temperature of the flowing gas is carefully controlled, which is not easy to achieve, the nebulization rate is not constant.

The mixture in the spray chamber is heated in some commercial instruments by infrared radiation. Again, as this reduces the overall droplet size, increases in sensitivity result. The system has not been universally adopted, however as an inherently simple component is thereby made more complex, and the heating effect may not be consistent if sample and blank solutions are not run with complete continuity. Some workers have found too that heating of the carrier gas or of the spray chamber itself results in instability or drift in signal output and that there is a more pronounced tendency to carry-over or 'memory' from one sample to the next.

Impact bead and counter-flow devices

Two forms of mechanical device have been introduced into the spray chamber in order to utilize more effectively the momentum of the droplets themselves to produce a comminuting action. In one, a bead or bar is placed close to the orifice of the nebulizer. The droplets, whose speed at this point is near sonic, are fragmented by impact and the mass of material vaporized in the flame can be increased by 50–100%. The material from which the bead or bar is made must necessarily be very inert chemically, and for most solutions, fused silica is satisfactory.

In a device described by Feldman[136] the oxidant/sample aerosol and fuel nozzles are placed in opposition within the spray chamber (Fig. 12). This also results in a high speed turbulence which produces a larger proportion of the smaller sized droplets. Maximum improvement in nebulization efficiency is obtained when the mass flow rates and velocities of the opposing streams are approximately equal. This produces a region of interaction midway between the nebulizer and fuel jets where the velocity of both streams is effectively reduced to zero and the liquid particles are thus further broken up by the high acceleration forces. Increases in sensitivity of two or three times have been obtained with this

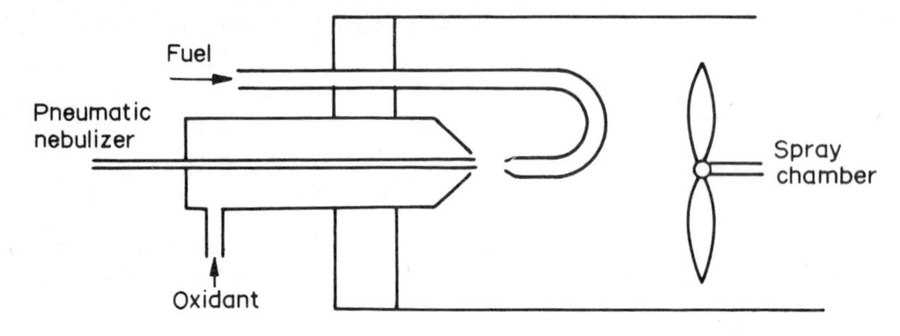

FIG. 12. Principle of counter flow nebulizer.

system, though it appears that the maximum improvement occurs fairly sharply at a critical distance between the jets which should lie on a common axis.

Ultrasonic nebulizer

A different approach to the problem of producing the largest proportion of volatilizable droplets from a solution is by the use of ultrasonics. While ultrasonic devices can attain higher nebulization efficiency they generally do so at a lower rate of nebulization. This means that they are able to convert 40% or more of small volumes of sample to a mist of useful droplet size, but are not able to produce this mist at a sufficient rate to give higher sensitivities when sample size is not a limiting factor.

Two basic ultrasonic systems are known—liquid coupled and vertical crystal. Ultrasound is produced by the piezo-electric effect on certain crystals, and the crystal vibrations have to be transmitted with the least possible loss of energy to the solution being nebulized. An example of the liquid coupled system was described by Stupar and Dawson.[378] Basically, the sample liquid, see Fig. 13, is held by a diaphragm in contact with a coupling medium (usually water) through which the ultrasonic vibrations are focussed by means of a concave crystal. The mist produced is swept away to the burner by a current of air. Frequencies used were 115 kHz and 70 kHz, and the majority of droplets were in the 17–21 μm diameter range.

A particular drawback of this device is that, while small samples can be handled, they cannot be fed in continuously as with the standard pneumatic type. To

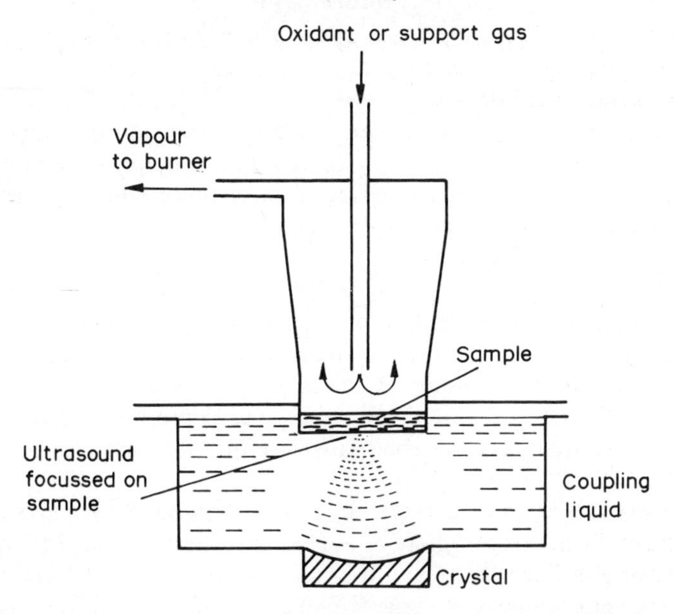

FIG. 13. Principle of liquid coupled ultrasonic nebulizer.

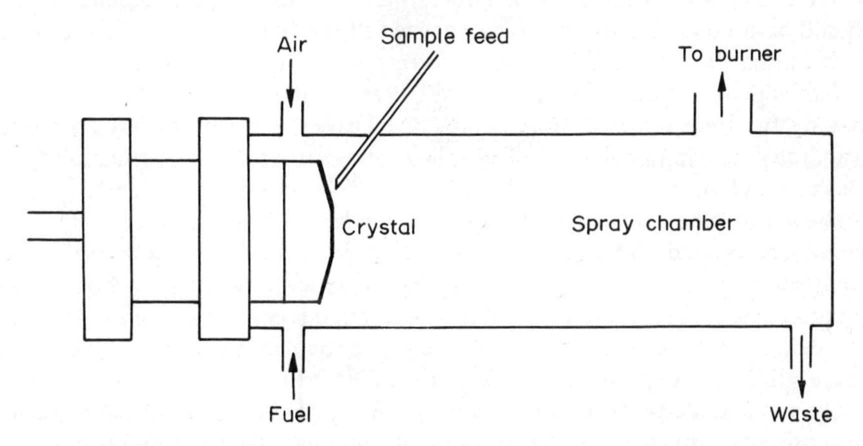

FIG. 14. Principle of vertical crystal direct ultrasonic nebulizer.

overcome this problem, the vertical crystal nebulizer was devised. Here the sample solution is injected continuously at a constant rate directly on to the vertical face of the vibrating crystal. The crystal must therefore be chemically inert and quartz is best in this respect. Stupar and Dawson[379] have also described the modification of a medical inhalation therapy instrument which employs a metal disc of resonant thickness as the nebulizing surface (Fig. 14). Nebulization efficiency of this system is said to be 50% at a sample flow rate of 0.1 ml/min and falls off to 24% at 3 ml/min.

These workers' conclusion was that the stability and reproducibility of the absorbance values obtained with the pneumatic nebulizer are better than with an ultrasonic system, the better efficiency of the latter being offset to some extent by its slightly erratic perfomance. They also suggest that, unless higher frequencies can be employed (e.g. 500 kHz or more) with the corresponding power to generate fine mist with droplet sizes not greater than 5 μm at sample flow rates of up to 5 ml per minute, the sensitivities and even interference effects are likely to be worse than with the pneumatic nebulizer.

Spray chamber

In addition to allowing the selection of sample mist of the wanted droplet size, the spray chamber also allows the sample mist, oxidant and fuel gases to become thoroughly mixed before passing on to the burner. Auxiliary oxidant, that is, further oxidant which is required to support the flame but which is over and above that required to make the nebulizer work efficiently, may also be introduced at this point.

The gases are mixed and the larger droplets settled out by imparting a rotary motion to them. Either one or more of the gas inlets and nebulizer are positioned tangentially, or else fixed vanes are inserted as shown in Fig. 9. One tube leads away from the spray chamber to take the mixed gases to the burner, and another to remove deposited liquid to waste.

If the nebulizer and spray chamber functions are properly matched, there should be no droplets whatsoever deposited in the burner tube or in the burner itself during the functioning of the instrument. The correct mist is virtually dry and should not condense on any surface. One school of thought even suggests that by the time the mist has reached the burner, the solvent has completely evaporated, leaving small suspended solid particles in a partially saturated vapour. This may well be true in view of the possible surface activity of such small droplets.

The waste tube leads first to a syphon, the head of which prevents the gases from escaping and also ensures a constant small excess of pressure in the spray chamber. For combustion mixtures using air as the oxidant a syphon head of about 2 cm is adequate, but for mixtures using nitrous oxide, this must be increased to 4 or 5 cm otherwise the back pressure at the smaller burner slot will cause the gas to bubble out with consequent loss of flame stability.

The size and shape of the spray chamber, within limits, should have little effect on sensitivity provided that it is working at optimum efficiency, and this is borne out in practice. The removal of the flow vanes, if these are employed, may result in an apparent small increase in sensitivity, but this will almost certainly be accompanied by an increase in flame noise with subsequent worsening of detection limits.

As the gases normally passing through the spray chamber constitute an explosive mixture within the confined space, all spray chambers should be provided with a safety device in case the flame burns back from the burner head. This may take the form of a rupturing membrane, or the ends of the spray chamber may simply consist of push-fit rubber bungs. Explosions of nitrous oxide and acetylene mixtures being somewhat more forcible than air acetylene, it is also a wise precaution to place a metal cage around the spray chamber, and restrain the burner head from flying off, if this also is a push-fit, by anchoring with a strong wire loop.

Burners

Burners are designed specifically for the various combustion gas mixtures employed. The major consideration is that the flame propagation velocity should not be greater than the gas velocity through the burner slot, otherwise the flame may flash back down the burner stem and into the spray chamber, with possibly disastrous results. This is often modestly referred to as a 'blow-back'. The flame is stabilized also by what Gaydon and Wolfard[152] called the quenching effects at the walls of the burner slot or holes. This is the cooling effect of the walls on the reacting gases. The quenching distance for a pair of parallel plates and the quenching diameter for a row or pattern of holes is that which will allow a given flame to be just held at the mouth of the burner. Quenching diameter is normally greater than quenching distance.

In addition to the heat-quenching effect of the walls of the burner slot or holes, the latter also exert a drag or friction on the moving gas. Figs. 15 (i) and (ii) represent the velocity V of gas in the burner orifices, in which the distances

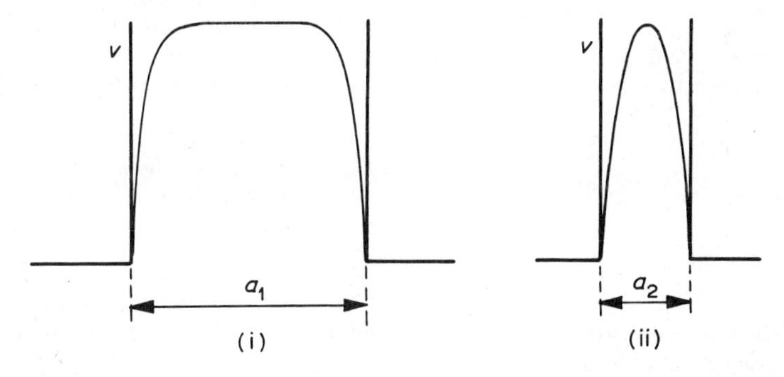

FIG. 15. (i) Gas velocity across wide burner slot. (ii) Gas velocity across narrow burner slot.

between the walls are a_1 and a_2 where $a_1 > a_2$. Thus the total gas flow, which is proportional to $\int_0^a V da$, the area under the curve, is not proportional to a. Hence flow rate is not proportional to slot area, and the smaller the area, the lower is the flow rate in proportion.

The effective burner mouth thus depends on the actual area and the shape and size of the holes or slots. In particular, for two burners of the same area, one consisting of holes will have a smaller effective area than a single slot; one consisting of two or more narrow slots will have a smaller effective area than a single wide slot; and a burner consisting of a long narrow slot will have a smaller effective area than one half as long but twice as wide.

The general form of burners for atomic absorption provides laminar flow and a long path length flame. In a laminar flow burner the direction of flow is generally perpendicular to the flame front. In turbulent flow, the direction is not well defined. Although it may appear that the longer the path length, the higher the absorbance will be, this is true only insofar as the effective slot area allows a given gas flow rate to be maintained. Higher absorbances are therefore produced by limiting the slot width, so that the absorbing species are concentrated in the narrow light envelope. This principle is illustrated by the direct substitution of a 5 cm path length 'nitrous oxide' burner for a 10 cm air acetylene burner for an air acetylene flame. The absorbance values are reduced by only about 10% due to the effect of the smaller slot area on the flow rates.

While it is always advisable to use the correct burner for a given gas mixture, it is sometimes possible to use a burner for a slower mixture than that for which it was designed, but it is imperative that it be not used for a *faster* burning mixture, otherwise a blow-back will occur.

In Table 3 the most common types of burner presently used with atomic absorption spectrometers are listed together with the flames they will support. In general,

TABLE 3.

Burner	Slot length	Gas mixture supported
Air acetylene	10 cm	air acetylene
		air hydrogen
		argon hydrogen (see p. 35)
		N_2O propane
Air propane	10 cm	air propane
Multislot ('Boling burner')	10 cm	air acetylene (with auxiliary air)
		air propane
		air butane
		argon hydrogen (see p. 35)
N_2O acetylene	5 cm	N_2O acetylene
		air acetylene

Meker-type emission burners may only be used for the flame they are intended for. It is, of course, a wise precaution to consult the manufacturers of any burner before using it for a different flame from that intended.

Multislot burner

A limitation of most single slot burners, or burners where the slot is replaced by a single row of circular holes, is that the maximum concentration of dissolved solids in the samples to be atomized may be of the order of only 2%. This applies particularly to burners with slots of small dimensions, e.g. air acetylene and nitrous oxide acetylene. Solid solute tends to build up on the underside of the jaws more or less quickly, producing a drift in the flame characteristics.

The effect was first overcome in the air acetylene flame by Boling[66] using a triple-slot burner and a higher throughput of oxidant gas. Multislot burners provide a smaller quenching distance than a single slot burner of the same total slot area, but need a greater gas flow rate than standard single slot burners, the amount of oxidant required usually being more than that which can pass through the nebulizer. An auxiliary supply of air, bypassing the nebulizer, is then introduced into the spray chamber, mixed and passed to the burner. Multislot burners can usually allow up to 12% of dissolved solids to pass evenly to the flame without fear of build-up on the burner jaws.

For some elements, the sensitivities produced in the multislot burner are worse than in the single slot, and for some other elements they are better. Two opposing effects appear to be in operation. The species being measured is effectively diluted by the higher volumes of combusting gases and the increase in equivalent slot width. But this can be more than compensated for by the increase in flame temperature of the central slot which is protected by the two outer slots. This applies particularly to those elements which are incompletely atomized at the temperature of the single slot air acetylene flame or whose degree of atomization is easily decreased in the presence of entrained air.

Flames in normal use

Air Propane. This flame is now rapidly losing favour, and many of the interference effects reported in the early literature were simply a result of its comparatively low temperature. It is much less stiff than the air acetylene flame and is liable to distortion by draughts or 'flicker'. It is therefore not recommended for instruments where the flame compartment is not enclosed. For those elements which do show an improved sensitivity, this is obtained only when the flame is completely non-luminous. Under these conditions the flow rate of propane may be as low as 300 or 400 cm^3/min with some instruments where the real air flow rate is 8 l/min.

The lean air propane flame has a very small blue inner cone which may tend to lift off the burner head. The stoichiometric flame is stable and just not luminous, while the rich flame is luminous and usually tends to be 'floppy'.

Air propane flames may be better sustained on a multislot burner, particularly if organic solvents are in use. In some countries propane may not be available – only propane butane mixtures or butane gas. Butane has roughly the same analytical characteristics as propane, but the flame will be successfully maintained only on a multislot burner.

Air Acetylene. Perhaps the most generally useful flame of all, this can be supported on the nitrous oxide acetylene and multislot burners in addition to its own burner. For most elements, best sensitivity is obtained with the stoichiometric flame, though others, notably chromium, molybdenum and tin, show best sensitivity when the flame is very rich and luminous. Under such conditions, increased noise, originating from the flame, occurs in the output signal.

The four types of air acetylene flame listed in Table 2 can be distinguished visually. The lean flame is very stiff with small inner blue cones. The stoichiometric flame is stiff with blue cones 2–4 mm in height, but is just not luminous. The blue cones can still be distinguished in the luminous flame, but tend to disappear in the rich flame, which is luminous almost to the point where sooty smoke appears from the top.

It may be useful to remember that when acetylene is taken from cylinders or tanks, these are filled with kieselguhr and acetone to dissolve the acetylene. The cylinder must therefore always be used with the valve uppermost, and it is wisest to discard the last 20% of the acetylene capacity as this is usually contaminated with acetone and the flame characteristics become different.

Nitrous Oxide Acetylene. Although this flame can be burned on a 10 cm burner slot, the optimum length for most metals is 5 cm, as this incurs less cooling effect and improves sensitivity. The burner slot size is thus usually 5 cm $\times$ 0.45 mm. The burner top itself is massive and either water- or air-cooled in order to avoid the possibilities of pre-ignition and blow-back through excessive heat at the burner slot.

The flame produced in this standard type of burner is characterized by three distinct regions: the primary reaction zone, about 2–3 mm high and whitish-blue in colour; the red interconal zone which takes the appearance of a red feather (by which description it is usually known) varying in height from 0–30 mm depending on the oxidant/fuel ratio; and a blue secondary diffusion zone. Kirkbright et al.[205] have shown that the red feather, where the highest temperature is to be found, gives strong CN and NH band emission, and have suggested that the high degree of atomization of some elements in this flame is at least in part due to the more complex reactions occurring with these species and not simply to the action of atomic or incandescent carbon as is often assumed for the air acetylene flame.

In the lean flame, the red feather is less than 10 mm in height, but this increases to 10–20 mm in the stoichiometric flame, which is just not luminous. In the luminous flame, the red feather again increases in height, and the top is generally lost in the brilliant luminosity.

With flat-topped burners and an acetylene gas flow rate of about 4 l/min, a glowing deposit of carbon tends to form along the edges of the slot, altering the flame characteristics. This can be removed without extinguishing the flame by pushing it off with a (preferably old) screwdriver or similar implement. This must be done quickly so that the implement used does not melt, but care must also be taken to avoid pushing incandescent particles down through the burner slot. The deposit is not formed at higher acetylene flow rates (e.g. greater than 5 l/min). A burner designed to avoid this nuisance has lipped slot edges. Here either the deposit has no room to lodge or the entrained air quickly oxidizes it.

The high gas flow rates have in the past caused operators to show some reluctance to ignite this flame directly with a taper or lighter. It is important to remember that blow backs will not occur as long as the flow rate through the burner slot is greater than the flame propagation rate. The critical flow rate for this gas mixture with the slot size given above is about 2.5 l/min. With individual gas flows of 10 and 3 l/min of oxidant and fuel respectively there is a very large safety margin. Indeed there is still a safety margin if either of the gases fails completely. However, some commercial instruments are fitted with self-ignition systems, and one allows the flame to be lit on air acetylene, with a solenoid operated valve to turn to nitrous oxide. This system is also useful as it enables rapid interchange between the flames when several elements are to be determined, the nitrous oxide acetylene burner always remaining in position.

Nitrous Oxide Propane. Although this flame is not now considered to be one of the more important in atomic absorption analysis, it has some useful features and a few details are therefore recorded in Table 2. Precautions advised for nitrous oxide acetylene should be observed. The flame is easily controlled and there is no tendency to the deposition of carbon along the edges of the burner slot.

Separated flames

The laminar premixed flames described so far always exhibit three distinct reaction zones. After the preheat zone, these are: (i) the primary reaction zone;

(ii) the interconal zone, which is small in stoichiometric flames but increases as the gas mixture becomes richer in fuel. The interconal zone is the most productive of free atoms for those elements which form stable monoxides; (iii) the secondary reaction, or diffusion zone sometimes called the 'plume'. These were separated as long ago as 1891 by Teclu,[387] and more recently Mavrodineanu[258] has made a study of the molecular reactions taking place in each zone. In the primary zone of the air acetylene flame inside the flame and above the blue region of unburnt gases, the products mainly formed are carbon monoxide, hydrogen and

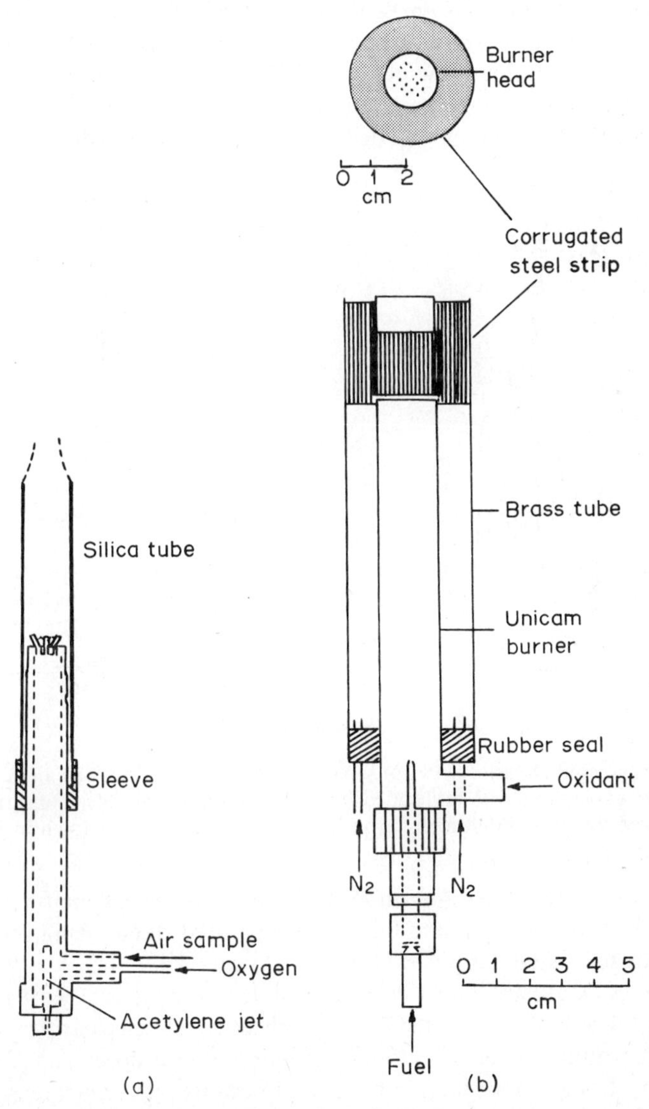

FIG. 16. (a) Burner arrangement for separated air acetylene flame employing silica separator tube. (b) Burner arrangement for separated air acetylene flame employing nitrogen shielding gas.

water. The secondary zone, a cooler outer mantle, is the diffusion flame in which these gases are burnt in the atmospheric oxygen to carbon dioxide and water.

Kirkbright and West[212] have examined the properties of the primary and interconal zones of the air acetylene and nitrous oxide acetylene flames for use in atomic absorption and fluorescence. Separation of the secondary zone was achieved mechanically in the first instance by supporting a silica tube on the burner stem with its open top about 7–10 cm above the circular Meker head. Lean gas mixtures tend to produce turbulent flames with this arrangement, and rich mixtures lead to the deposition of carbon on the walls of the tube, making observation uncertain. A better system is to surround the flame with a curtain of inert gas such as nitrogen or argon. The means whereby this is achieved is illustrated in Fig. 16. Although this shows a circular Meker burner, the same kind of device can be placed around a long path length atomic absorption burner. The potential value of separated flames can be deduced from Fig. 17 in which the background

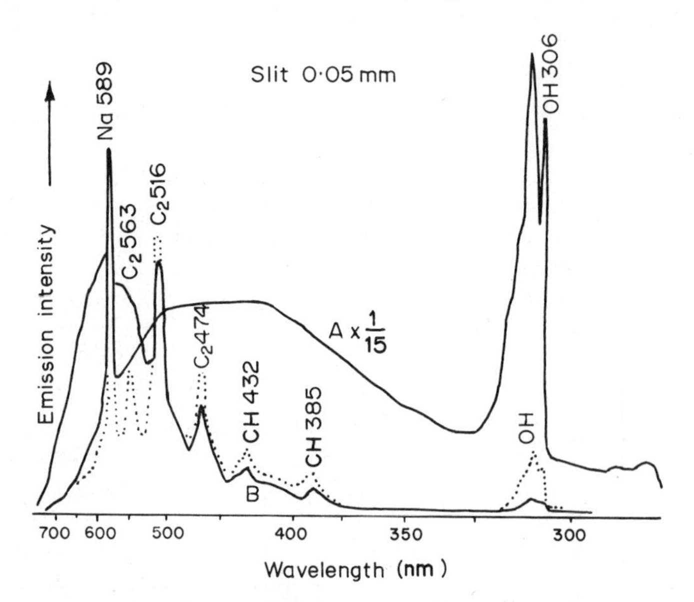

FIG. 17. Background emission from unseparated (A) and separated (B) air acetylene flame. The ordinate of A should be multiplied by 15 to compare absolute intensities. The effect of decreased fuel flow rate on emission from separated flame is shown by broken line.

emission from unseparated (A) and separated (B) air acetylene flames are superimposed, the intensity ordinate for the unseparated flame being on a scale 1/15 of that of the separated flame.

The OH and CO band emission in the interconal zone is at least two orders of magnitude lower in the separated flame. Such an effect will clearly help to improve signal/background ratios in atomic emission and fluorescence, and also in absorption, though more indirectly. This is because the very intense band emissions increase the photomultiplier shot noise (see p. 47). In addition they may saturate the detector or contribute markedly to flame noise by their component of amplifier modulation frequency.

Separation of the diffusion zone of the nitrous oxide acetylene flame was described by Kirkbright, Semb and West[207] and considerable improvements for a number of elements in flame emission spectrometric measurements are quoted. Indeed, detection limits for aluminium, beryllium, molybdenum, titanium and vanadium appear to be as good as or better than those expected by atomic absorption.

In atomic absorption, separation of the air acetylene flame would seem to have the most benefit for the determination of elements whose resonance lines occur at low wavelengths, especially arsenic and selenium, where the measurements are made below 200 nm. At such wavelengths, absorption by flame gases in the separated flame is much less than in the normal flame and so better sensitivities and detection limits are obtained. Separation of the nitrous oxide flame maintains the reducing properties of the interconal zone flame gases by preventing the diffusion of atmospheric oxygen. Better sensitivities are thus obtained for some of the elements, such as aluminium, boron and silicon, which form highly refractory oxides.

The possible advantages of using separated flames in analytical atomic spectroscopy were discussed in some detail by Cresser and Keliher.[95]

Diffusion flames

The high degree of absorption of radiation of low wavelength by hydrocarbon flames has led some workers to examine flames based upon hydrogen. These are much more transparent at low wavelengths. In diffusion flames, the fuel is surface-mixed with ambient air, and not pre-mixed with oxidant. Such flames are usually cooler than the corresponding pre-mixed flames,[466] but allow a very safe, non-explosive system to be devised.

Hydrogen diffusion flames are supported on a standard air acetylene or multislot burner and the hydrogen is usually diluted with an inert gas such as nitrogen or argon. When this type of flame is used with standard atomic absorption equipment no modification is required as the diluent gas may be used for nebulization. The apparatus is set up as follows: the nitrogen or argon supply is connected to the 'oxidant' inlet of the instrument and the hydrogen supply to the fuel inlet. Extra care must be taken to ensure that the spray chamber and associated tubes are leak free because of the high diffusibility of hydrogen gas. The spray chamber is flushed for a few seconds with the hydrogen before ignition, the hydrogen flow is adjusted to about 1.5 l/min and then ignited. After ignition, the diluent gas is turned on. A convenient flow rate for argon is 5 l/min if an air acetylene burner head is being used, but 6 l/min or more will be required with a multislot burner. If necessary the 'auxiliary oxidant supply' system can be utilized to increase the argon flow rate to the burner. Higher flow rates can also be used for the hydrogen.

The type of flame described here was first used by Kahn.[196] Menis and Rains[268] used a total consumption burner with the same gas mixture and Nakahara, Munemori and Musha[284] used an air hydrogen flame.

The use of argon has one particular advantage over nitrogen. Because of its lower heat capacity, the temperature of the flame is higher, and better sensitivity figures generally result. Incidentally the quenching effect is lower than that of nitrogen, and thus it is a much better flame to use in atomic fluorescence. A disadvantage is its relatively higher cost.

THE OPTICAL SYSTEM

As will be seen from Fig. 8 the optical system of an atomic absorption spectrometer has four main constituent parts. These are:

(i) the primary radiation source;
(ii) the pre-slit optics by means of which the source radiation is first focussed in the centre of the flame position, then brought to a second focus at the entrance slit of the monochromator;
(iii) the monochromator;
(iv) the detector.

The parts are discussed separately but they must also be considered as an integral unit for, in a properly designed instrument, best performance can only be obtained by careful matching of all optical parameters.

Primary radiation source

In Chapter 2, Resonance Radiation, it was shown that the primary radiation source must emit a sharp resonance line spectrum. Ideally a half-width of 0.001 nm should be provided. Appreciable absorption, though not the maximum possible, is still obtained with emission lines of 0.01 nm half-width. Thermally excited spectra, and arc and spark spectra, contain lines that are broadened by Doppler and pressure effects. Arc and spark spectra are also subject to considerable electrical and magnetic field broadening. They are therefore unsuitable as the primary radiation source in atomic absorption.

Vapour discharge tubes were the first sources to be used, as they are readily available for some of the volatile common metallic elements. Undoubtedly the best source is the sealed-off hollow cathode lamp, which has undergone considerable development and improvement in intensity and reliability since atomic absorption became popular. More recently, the possibilities of microwave excited electrodeless discharges have been carefully investigated.

Vapour discharge tubes

Vapour discharge lamps consist of a glass or silica tube containing an inert gas at a pressure of several torr and a quantity of the metal whose spectrum is to be produced. Oxide-coated electrodes are sealed in. They can be run on a.c. or d.c. If a.c., they are controlled with a variable transformer and ammeter. When first switched on, a gas discharge takes place which warms and vaporizes the sealed-in metal. The metal vapour then takes over the discharge and the radiation consists almost entirely of the metal spectrum. Such lamps should be run at currents considerably below those recommended by the manufacturers in order to minimize self-reversal of the resonance lines. For example a current of less than 0.5 A is desirable instead of the normal rating of 1 A or more.

This type of lamp is available for sodium, potassium, zinc, cadmium, mercury and thallium, and these are manufactured by Osram and Philips. Elenbaas and Riemans[129] have given details about their construction and operation. Lamps which do not have a quartz outer envelope may require a hole to be cut in the protective glass envelope, or the envelope removed altogether in order to pass the resonance lines in the ultraviolet.

Vapour discharge lamps have the advantage of much higher intensity than hollow cathode lamps. This means that better ultimate signal to noise ratios should be obtained, and therefore better detection limits, but at the required low currents, these lamps tend to be more unstable, and the advantage is to a large extent ruled out. However, in atomic fluorescence measurements, where the sensitivity is a direct function of source intensity, and where excitation line widths are less important, much better analytical performance is achieved.

In atomic absorption, modern hollow cathode lamps are undoubtedly better from the point of view of stability and line width for cadmium, zinc, thallium and even mercury. Good hollow cathode lamps are now also available for sodium and potassium, and the use of vapour discharge lamps appears to be dying out.

Hollow cathode lamps

The hollow cathode discharge has been known for many years to spectroscopists as a fine line source, and indeed one capable of producing spectra where the fine structure could be studied.[396] It has also been used as the excitation source for microsamples in emission spectroscopy. In these applications the source is, of necessity, demountable, and indeed in the earliest atomic absorption work, it was also used in demountable form.

Demountable Hollow Cathode Lamps. These allow the cathodes to be changed at will, and hence one lamp system can be used for any element (or combination of elements if suitable alloy cathodes are available). After each change of cathode, however, the lamp has to be purged with fill gas and then pumped down to the operating gas pressure. A refinement in some systems is that the fill gas is kept circulating through a cleaning-up section. In such cases, the cathode never becomes poisoned, and the atom cloud which normally forms in front of the cathode, increasing the self-absorption of radiation, is effectively removed. Stable high performance is therefore possible. A demountable hollow cathode lamp system is, however, costly to set up, as it requires high vacuum techniques, and the time taken to change from one element to another is too long to be considered in routine analysis. A system ideally suited to research requirements has been described by Elwell and Gidley.[1]

For the above reasons, therefore, it is the sealed-off type of hollow cathode lamp which is in common use as the primary radiation source at the present time. The hollow cathode discharge is in fact a low pressure discharge ('Geissler' gas discharge tube) with a special geometry. If the cathode is made hollow in shape, and the gas pressure reduced to the correct value, the discharge takes place entirely in the hollow cathode, when the radiation emitted becomes rich in the spectrum of the cathode material in addition to the spectrum of the fill gas. The pressure is somewhat higher than for a gas discharge tube.

Typically, the internal diameter of the cathode would be from 2–5 mm, the fill gas argon or neon, and the operating pressure between 4 and 10 torr. This is contained within a glass envelope through which the cathode and a tungsten wire anode are sealed. The whole is energized with a potential of about 300 V, and currents of 4 mA up to 50 mA or more may be passed, depending on the element being excited.

The choice of fill gas depends primarily upon two factors. Firstly, the discharge emission lines of the gas itself must not coincide with the resonance lines of the elements to be measured. Secondly, the relative ionization potentials of the fill gas and cathode metal must be taken into account. The ionization potential of neon is much higher than that of argon and many metals, and hence will tend to increase the proportion of 'spark' lines produced, as excitation is caused by collisions of the second kind. Neon is therefore normally used with elements of high ionization potential. Other considerations are that argon, being heavier, has a more efficient sputtering action, and also 'cleans up' (adsorbs on sputtered metal films) less quickly; on the other hand, the spectrum of neon is much less rich in lines than that of argon, and hence neon is probably better when the lamps are used with instruments of lower dispersion.

Sealed-off Hollow Cathode Lamps. The earliest sealed-off lamps (see Fig. 18) consisted of a glass tube through which the electrodes were sealed, with an optical window of glass or silica (depending on the wavelength of the wanted

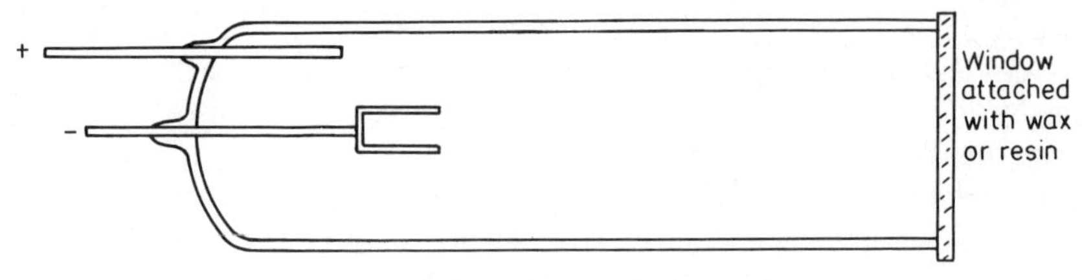

FIG. 18. Early hollow cathode lamp.

resonance line) attached, usually with a thermosetting resin or vacuum wax. The cathode normally had an internal diameter of 10 mm. The components of this device had to be cleaned up by baking before finally pumping down to operating pressure. Because of the wax sealing, very efficient cleaning was not possible, and the fill gas and cathode became poisoned more or less quickly. This led to a reputation of short life for early sealed-off lamps, but this has fortunately now been dispelled.

The move towards all-fused construction and modified geometry over the past five years has now led to a lamp which provides much higher intensity over a long useful life. Most manufacturers guarantee their lamps for 1000 hours and they usually run for twice this length of time. No more than 2 or 3% of reputable lamps fail under this guarantee, such is their present-day reliability.

An internal cathode diameter of 2 mm is now much more common (Fig. 19) as this concentrates the energy of discharge on a smaller area and thus produces a much higher intensity. Also, the energy appears to be dissipated to a greater extent in the resonance lines, giving these a higher intensity in relation to the rest of the metal spectrum and the gas lines. This clearly results in better analytical performance. A mica shield helps to prevent outward spread of the discharge, and for the same reason some lamps are provided with a ring anode. However, the anode shape does not appear to be of major importance. A higher gas pressure

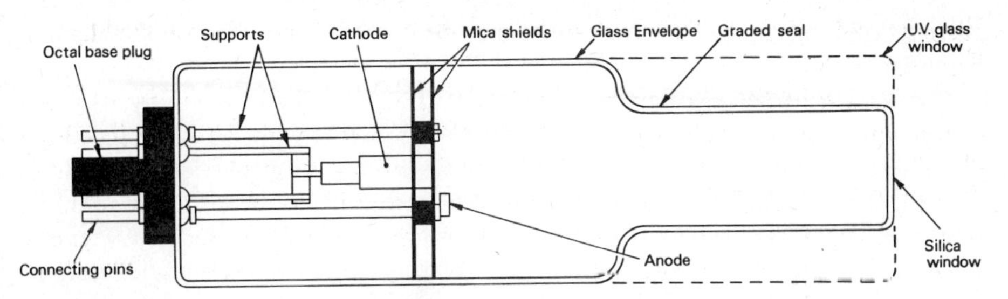

FIG. 19. Diagram of modern hollow cathode source.

is used – up to 10 torr – and this both helps to maintain the discharge inside the cup. and takes longer to be 'cleaned up' in use. Fig. 20 shows the relationship between operating current and gas pressure for hollow cathode lamps in general.

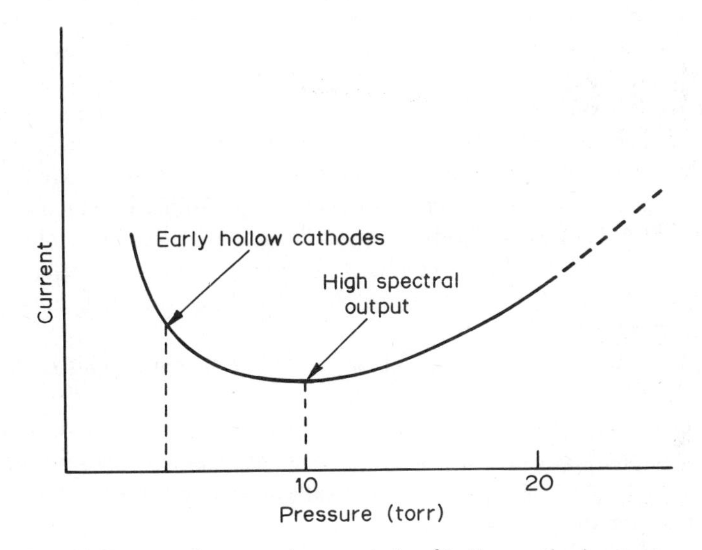

FIG. 20. Pressure/current characteristic of hollow cathode sources.

This shows that this modern type of lamp – called by several manufacturers a 'High Spectral Output' lamp can be run to advantage at a lower current than the corresponding early lamp. There are thus several factors contributing to improving the length of working life and analytical performance.

In the all-fused construction of these lamps, ultraviolet transmitting-glass is used for the body and the optical window where the resonance line has a wavelength higher than 250 nm. Below this wavelength the transmission of ultraviolet glass falls off gradually (e.g. 70% at 240 nm). A silica window must be used instead, and this is attached to the body by means of a graded seal.

Boosted Output Lamps. This is a type of hollow cathode lamp, introduced by Sullivan and Walsh[381] in which the cloud of atoms normally formed in front of the cathode is itself excited by a secondary discharge. In this way these atoms are made to *contribute* to the intensity of the resonance radiation rather than diminish it by absorption. The secondary discharge is of a low voltage but higher current (300–400 mA). The auxiliary electrodes are almost entirely enclosed in glass sleeves but openings are provided to position the discharge correctly.

When introduced, these lamps (known as 'High intensity' and 'High brightness' by two different manufacturers) improved signal-to-noise ratios by virtue of their increased intensity, and also straightened calibration curves because the resonance lines given were less self-absorbed before reaching the flame. Their higher degree of complexity, however, seemed to make them less reliable and shorter lived, and they appear to have no real advantage over the high spectral output lamps described above.

Multi-element Lamps. The use of multi-element lamps, i.e. lamps whose cathodes contain more than one element, has been advocated, partly on the grounds of economy and partly for their convenience.

It is difficult to prove their economy, because when the lamp fails, one has lost all the elements. Also, for some reason which is not entirely apparent, such lamps usually cost more than those for a single element. It is clearly not economical to use a multi-element lamp when a single element only is being determined, as this is equivalent to wasting the life of the other elements.

Experience has shown that the different rates at which elements sputter cause them to lose their intensity and be lost in turn. The alloy from which the cathode is made should therefore contain the elements in concentrations which are proportional to their rates of sputtering under the conditions in which the lamp is to be used. Alternatively, cathodes formed by powder metallurgy or by discrete rings of individual metals have been used, with varying degrees of success. There is thus a considerable limitation on the elements which can be brought together in a single cathode.

The convenience of a multi-element lamp is without doubt. This is particularly so where a routine analysis for some three or four elements is regularly carried out and those elements can be obtained together in one cathode. 'Warm-up' and change-over times can then be reduced. However, the resonance line intensities of individual elements are somewhat diminished as a result of the elements being brought together in one cathode.

Hollow Cathode Lamp Currents. The supply to a hollow cathode lamp should be either current-stabilized or limited with a series resistor, as the current is almost independent of the applied voltage. A given lamp will operate over a fairly wide range of currents, but it is desirable to choose the best operating current for the particular lamp and analysis in hand. Some manufacturers label lamps with 'maximum' currents and others with 'maximum operating' currents. A maximum current should under no circumstances be exceeded as irreparable damage may be done. High currents, and currents above 'maximum operating' values will result in excessive line broadening. It is therefore nearly always desireable to run a lamp at a current value less than that given on the envelope. Many users prefer to run the lamp at the lowest current that gives adequate stability and freedom from intensity drifts. This improves the sensitivity because of minimal line broadening, and increases lamp life. With modern instruments it should be unnecessary to run lamps at higher currents with the object of reducing the amplifier gain to lower the electronic noise levels.

For a single beam instrument the lamp 'noise' should be less than 0.2% and intensity drift less than 2% per hour. For a double-beam instrument, in which variations in intensity are to a large extent compensated, these values may be relaxed.

Lamps need a short warm up period before use, the length of which depends on the type of lamp and the element. Periods of between 5 and 20 minutes are normal. During this time the lamp intensity drifts to its equilibrium value, sometimes gradually upwards, sometimes going first above and then slowly downwards. During this time, also, the atom cloud mentioned earlier attains its equilibrium density so that self-absorption of resonance radiation, and hence the line-shape absorptivity coefficient of the line, also show drift towards a stable value. These effects are discussed again in relation to single-beam and double-beam optical systems.

Microwave-excited electrodeless discharge tubes

Some of the disadvantages of hollow cathode lamps, viz. comparatively low resonance line intensity, high cost, are said to be overcome by electrodeless discharge tubes. Again, such sources are not new, having been used for studies in spectral structure for some years, but their use as intense sources for atomic fluorescence has been pioneered by Winefordner and Staab[451] and as line sources for atomic absorption and fluorescence by Dagnall *et al.*[105–108]

The electrodeless discharge tube takes the form of a sealed, quartz tube, 3–8 cm in length, 1 cm or somewhat less in diameter, containing a few milligrams of a metal or volatile metal salt. The tube is filled with an inert gas at a pressure of a few torr which starts the discharge and helps maintain it with collisions of the second kind. A microwave-frequency electromagnetic field provides the excitation energy through a wave-guide cavity. Microwave frequencies have been found to be more efficient than radio frequencies as the excitation medium, and some workers have found that the tubes give a longer useful life at these higher frequencies. For much of the work carried out to the present time, the power source has been one of the comparatively inexpensive medical diathermy units (Electro-medical Supplies Ltd.) which provide power at about 2450 MHz up to 200 W. These can be connected to various wave-guide cavities and antennae.

Detailed instructions on the preparation of electrodeless discharge tubes have been published by Dagnall and West.[110] The material placed in the tube to generate the discharge must have a vapour pressure of about 1 mm at 200–400° C in order to function with the power unit and cavities mentioned. This is the factor which dictates whether the metallic element itself, or whether its chloride or iodide is to be used. In the case of the iodide, this may well be best formed by using a little of the metal and a small excess of elemental iodine. The inclusion of one or two milligrams of mercury – or even just saturated mercury vapour – is said to improve the reliability of 'striking' of the lamp by acting as a carrier of the initial discharge. It also tends to improve lamp life by preventing adsorption of the main element on the quartz tube.

The thermally uniform nature of the discharge and even a skin effect (concentrating the discharge near the tube walls) ensure that there is little or no self-reversal of the emitted resonance lines. However, the tube must be small enough to allow this condition to obtain without movement of vapour from cooler parts of the tube itself. There is thus an optimum size of tube for each element, which depends upon the vapour pressure of the material used and the nature of the cavity. It follows that a tube made to contain two elements can only be successful if the metals or compounds used have similar vapour pressures at the working temperature of the tube.

The cavities employed should be tunable and air-cooled. A reflected power meter may be used with some advantage to assist in tuning the cavity and to ensure that the reflected power itself is not sufficient to damage the magnetron. If the power is less than 75 W, however, visual tuning may be adequate.

When correctly prepared and run, electrodeless discharge tubes should emit only the principal lines in the spectrum of the main element, together with the iodine line 206.2 nm and several of the strongest mercury lines (assuming that the metal iodide is the active substance, and that mercury has been introduced either deliberately or incidentally by way of the vacuum pump). Lines may, of course, also originate from impurities in the substances used. Resonance line intensity of the wanted elements may well be ten to one hundred times greater than in hollow cathode lamps.

Dagnall claimed higher output stability for electrodeless discharge tubes than for hollow cathode lamps. A figure of $\pm 1\%$ was quoted.[110] Certainly very high stability for certain elements, particularly arsenic, has been demonstrated by some manufacturers who have also investigated this type of source. The fact seems to remain however that such stability and operating reliability is given by a comparatively small number of elements, in particular those with a high vapour pressure such as arsenic, antimony, bismuth, selenium and tellurium. It is many people's experience that a tube which operates perfectly on one occasion may give less than satisfactory results on another. While this may well be due to some difference in position within the cavity or variation in excitational parameters, such unpredictable behaviour renders some tubes unreliable in the hands of a routine analyst inexperienced in this particular art. It now appears however that better stability is obtainable at the expense of some intensity.

Tubes have been prepared for about fifty elements, but of these the five listed above are certainly the best. This is a happy coincidence because, probably for the same reasons of elemental volatility, these five also have a reputation for making the worst hollow cathode lamps. Electrodeless discharge lamps are thus a very useful tool in the hands of a research analyst, who can construct them as the need dictates, thus saving considerable expenditure on hollow cathode lamps.

At least one company markets these tubes with an improved stabilized power supply and undoubtedly, if others are able to improve their predictable performance in routine analysis, the electrodeless discharge tube might well rival the hollow cathode lamp in its position as the best resonance line source for atomic absorption analysis.

Pre-slit optics

The Optical Aperture. A major constraint in the design of the system is the long narrow flame. As much light from the source as possible must pass through the flame, otherwise lower absorption (and low sensitivities) than the maximum possible will be experienced. Ideally, the flame should fit the light envelope like a 'minimum volume' infrared gas absorption cell. Some attempts have been made to achieve this, without great success. The fact that the source is not a point but has finite dimensions tends, to some extent, to improve the situation because the light envelope is less conical than cylindrical in shape.

Because the source is focussed on to the monochromator slit, more energy will pass into the monochromator as the source image width (or diameter if, as is usual, the source is circular) approaches the slit width. A small source is thus preferred, and the small cathodes of modern high spectral output hollow cathode lamps are advantageous also in this respect.

A typical pre-slit optical diagram is given in Fig. 21 with the width of the light envelope much exaggerated. The long narrow flame is seen to influence the angle of acceptance at the monochromator slit, dictating the optical aperture of the monochromator and also of the source lamp itself. A narrow acceptance angle means that, for the purposes of atomic absorption measurements, a large monochromator aperture is unnecessary, and may even be an embarrassment as stray light levels (i.e. light not subjected to the absorption process in the flame) are then greater.

The design of multi-purpose instruments – which are to be used for flame emission or fluorescence as well as absorption – must therefore include a compromise, since in these, the detection limit in emission or fluorescence can be improved by making the flame fill a larger optical aperture.

It thus turns out that a monochromator aperture of f: 10 such as is normally used with ultraviolet spectrometers is suitable for atomic absorption requirements. A good flame emission spectrometer should have an aperture of f: 4 or thereabouts, and an instrument to be used for measuring flame fluorescence should have the same. Such a monochromator can be quite expensive but instruments designed to perform emission and/or fluorescence measurements as well as absorption usually compromise with an aperture of not smaller than f: 7.

In some early home made atomic absorption instruments the pre-slit optics consisted of two slit-shaped diaphragms, one at each end of the burner. The source lamp was placed as close to one of these as possible, and the monochromator slit close to the other. Such a system helps to reduce the amount of stray or unabsorbed light from the lamp falling on the detector, but it also seriously limits the total energy throughput. This results in good sensitivity values but poor detection limits.

Single and Double Beam Systems. The pre-slit optical system shown in Fig. 21 is typical of a single-beam instrument. As in any other single beam ultraviolet spectrometer, the light falling on the detector is proportional to the *transmission* of the 'sample' in the optical path,, and it is thus necessary to make a reading with and without the sample in order to obtain the *absorbance* which by Beer's law (see p. 77) is nearly a linear function of the concentration. Accurate absorbance values thus require a primary source of stable intensity. Stability in this context should, ideally, be absolute over the whole period during which measurements are taken so that the need to run reagent blanks or other reference solutions at frequent intervals is obviated. While modern hollow cathode lamps and good

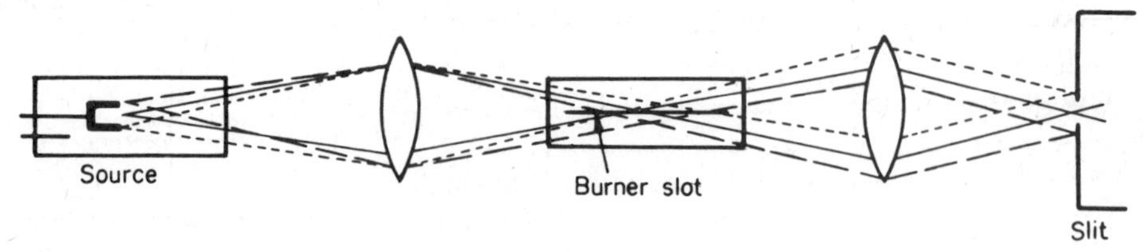

FIG. 21. Pre-slit optics, single beam, plan view.

microwave excited discharges certainly emit at the same intensity over long periods, no source is entirely without some intensity drift.

The effects of source variation can be overcome to a very large extent by employing double beam optics. In the most usual form of this device, the beam falls onto a rotating sector mirror before passing through the flame. This directs the beam alternately through the flame and along a path by-passing the flame at a frequency which may be 50 Hz or higher. Past the flame, the beams are recombined with a half-silvered mirror as shown in Fig. 22. At the detector, the

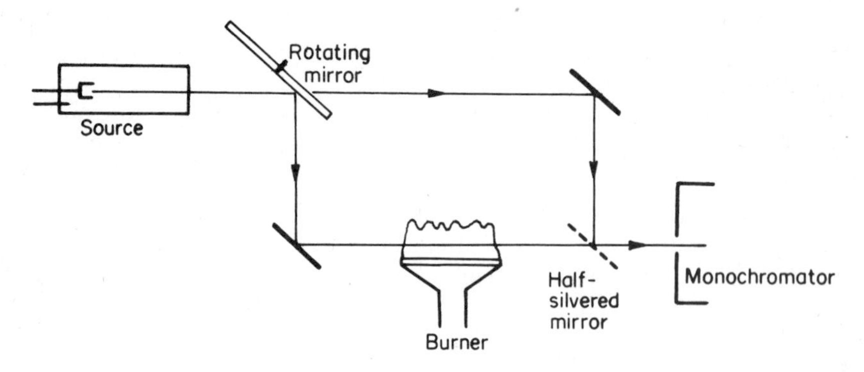

FIG. 22. Double beam optical system.

output signals corresponding to each beam are divided, amplified separately, and compared in a bridge circuit. The out-of-balance signal is then compensated by adjusting a potentiometer which can be calibrated in percentage absorption or, with suitable electronics, in absorbance.

A double beam atomic absorption spectrometer does not have all the advantages imputed to a double beam 'ultraviolet and visible' spectrometer. In the latter the reference beam passes through a cell which can be made to contain everything except the actual species being measured. Thus the true absorbance is obtained directly. In the atomic absorption instrument the reference beam does not pass through the flame. If it passed through a second flame, flames being *dynamic* systems, neither the chemical nor the noise characteristics could be guaranteed to be even similar, and certainly not identical. The normal double beam system, therefore, corrects only for variations in primary source intensity. It does not correct for any spurious absorption or scatter in the flame, nor will it allow for variations in the absorptivity coefficient of the primary radiation, which, as has been noted, occur particularly during the initial warm-up period of the primary source.

A further disadvantage of double beam optics is that, of necessity, at least 50% of the incident energy is wasted. Otherwise, the beams cannot be recombined without modulating the flame emission.

Modulated Sample Input. An interesting way of retaining most of the advantages of double beam systems is to modulate the aerosol supply to the flame. Antic-Jovanovic *et al.*[39] achieved this with two matched nebulizers and a fluid-switching technique. By dispensing with a spray chamber, modulation frequencies of up to 70 Hz were possible. With such a device, true background and flame noise effects are overcome by spraying the sample through one nebulizer and the blank solution through the other.

Monochromator

Resolution. With the ultimate resolution of the system depending on the spectral bandwidth of the resonance line emitted by the primary radiation source, the function of the monochromator is to isolate that resonance line from non-absorbing lines situated close to it in the source spectrum. Such lines may originate from the cathode metal or the lamp fill gas. The monochromator resolution should therefore be the best that will separate lines in the most complicated source spectrum to be employed, but as this will depend on the analysis to be undertaken, and to a great extent, the lamps to be used, no hard and fast rule can be given.

A monochromator of high resolution is not required when the alkali metals and perhaps calcium and magnesium are to be determined in biological samples, but for analysing metallic alloys, particularly those containing nickel and chromium, better resolving power is required. The problem is accentuated when source lamps emitting the spectra of more than one element are employed. A general purpose monochromator should thus be capable of separating two lines 0.1 nm apart or less when operating at minimum effective slit width. The better the quality of the monochromator, the smaller the slit width will be before further narrowing has no more effect on resolution. In very good instruments the minimum spectral band pass may be about 0.01nm.

A further function of the monochromator is to isolate the measured resonance line from molecular emission and other background continua which originate in the flame. To prevent all radiation from the flame reaching and 'saturating' the detector, the monochromator is always placed in the optical path after the flame.

With the aperture dictated by the limitation of the pre-slit optics, the maximum effective slit height is the diameter of the source image of the primary slits. Longer slits would simply allow unwanted radiation to enter the monochromator during absorption measurements.

Prism or Grating. The remaining choice lies between a prism or grating dispersing element. A glass prism is unlikely to be considered as this will not pass ultraviolet light. The essential difference in performance between prisms and gratings is that the dispersion given by prisms is high in the ultraviolet but decreases rapidly with increasing wavelength. For gratings, it is substantially constant throughout the spectrum, and depends on the number of grooves per unit width, the spectral order, and the focal length of the collimator.

Prisms can therefore be quite useful in atomic absorption work, as the majority of resonance lines occur in the ultraviolet region. At the other end of the spectrum, the alkali metal resonance lines are separated both from each other and from lines of other elements, though gas lines produced in hollow cathode lamps may cause stray light effects when these elements are determined in absorption. In general, however, all the light is transmitted by a prism in a single 'order' so the energy concentrated in a particular line is higher, and there is less stray light and other spurious reflections such as unwanted orders or ghosts.

Although still used in many excellent commercial spectrometers, prisms fail mainly on their lack of resolution in the range 240 nm upwards, particularly up to about 450 nm, within which region occur many important resonance lines of elements with complex spectra. Gratings are therefore becoming far more widely used, as good replicas are no longer expensive. The light energy passed falls off rapidly in the higher orders, so gratings blazed to give a maximum diffraction at a particular wavelength in either the first or second order will usually be adopted. In some instruments two gratings may be used to cover the whole range e.g. 180–400 nm and 400–860 nm, to maintain good resolution in the important lower wavelength region.

Detectors

The use of photographic plates and gas-filled photocells is now of historical and limited research interest only. Photomultiplier tubes are in almost universal use. The wide spectrum range to be covered (from the arsenic line at 193.7 nm to the caesium line at 852.1 nm) by a general purpose analytical instrument poses some problems of sensitivity, particularly towards the higher wavelengths.

The spectral sensitivity of photomultipliers depends primarily upon the photosensitive material used to coat the cathode. These materials are usually alloys of alkali metals with antimony, bismuth and/or silver. Most of these materials provide adequate output at wavelengths down to 190 nm provided the envelope material has adequate transmission. Caesium–antimony cathodes operate well up to 500 nm and the output falls rather steeply to around 760 nm, where there may or may not be sufficient sensitivity to detect the potassium and rubidium lines (766 nm and 780 nm). Photomultipliers of this type may well have to be selected for adequate performance. Trialkali cathodes, antimony–sodium–potassium–caesium, respond well enough up to 850 nm, and can usually be relied upon if caesium is to be detected. This type of photomultiplier is now manufactured in a side-window version which is only slightly inferior to the end-window tubes hitherto available and costing about the same as caesium–antimony tubes. More recently, gallium arsenide cathodes have become available, and their response at high wavelengths is a considerable improvement upon that of other types.

Typical spectral sensitivity curves are given in Fig. 23. These vary somewhat with different manufacturers and even different batches. Principal manufacturers are Philips, EMI, RCA and Hamamatsu.

Important characteristics of a particular photomultiplier from the point of view of atomic absorption are its dark current and noise. An amplification factor

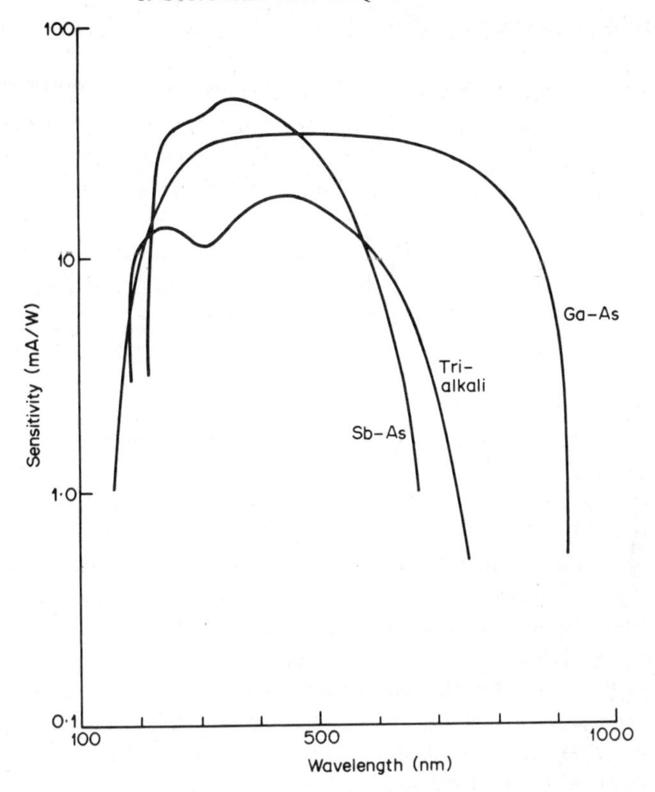

FIG. 23. Photomultiplier sensitivity curves.

within the tube itself of 10^6 or greater is commonly achieved by increasing the voltage applied over the dynode system. This also results in an increase of dark current and the noise component of the dark current (known at 'dark noise'). Also increased by increasing dynode voltage is the 'shot' noise, that is, the actual statistical variations in output caused by the electron showers generated between dynodes. Shot noise is proportional to the square root of the intensity of the radiation falling on the photocathode. To a point therefore, photomultipliers allow a relatively large almost noiseless gain factor, and it is desirable for the operator to have control of the overall instrumental gain both by the photomultiplier gain and the amplifier gain. In this way the optimum condition for minimum noise can be selected.

The gain g of a photomultiplier increases exponentially with the interdynode voltage v. The relationship is $g = kv^{0.7n}$ where n is the number of dynodes. It is therefore most important that the high voltage supply should be highly stabilized – e.g. to 0.05% or better – and that there should be a minimum of switching in the dynode circuit. In one commercial instrument, for example, two photomultiplier 'gain' positions are available to the operator with a factor of 100 between them. This together with the associated amplifier gain position gives a total amplification factor range of more than 10000.

READ-OUT SYSTEMS

Up to this point the component sections of an atomic absorption spectrometer have been discussed in some detail, as in the first place they are common to virtually all available instruments, and in the second place, an understanding of their function and functioning is essential by the practising analyst. The way in which the instrumental readings are presented probably most distinguishes one instrument from another. It is therefore of interest to know what the readout section of the instrument can be expected to do, but beyond the scope of this book to explain the many ways of doing it.

Modulation

The signal received by the detector consists of the resonance radiation (attenuated or not by absorption in the flame) and emission from the flame. Flame emission may include both resonance emission of the wavelength at which absorption is being measured, and molecular band emission and scatter from small particles in the flame, both of which are likely to extend beyond the monochromator bandpass. Only the resonance radiation originating from the source lamp is wanted. Other radiation falling on the detector diminishes the value of the absorbance that can be recorded. The lamp output is therefore 'coded' by modulation and the post-detector amplifier is tuned to the same modulation frequency, thus effectively preventing the continuous signal from the flame from being recorded. The source can be modulated either by using an a.c. supply current, or by interposing a synchronous chopper in the beam before the flame. It should be remembered however, that the total light signal still falls on the detector, and that if the unwanted part is very high, the photomultiplier can become 'saturated' and thus give a difference signal which is not proportional to the absorption being measured. Even before this happens a substantial contribution is made to shot noise, which, it will be remembered, is dependent on the total incident light intensity. A further problem arises if the continuous signal contains a high noise component, such as would be the case with a highly emissive or luminous flame. The noise will almost certainly contain a contributing frequency the same as the modulation frequency. There will thus be a noise breakthrough which will obscure the wanted signal. This is impossible to remove electronically, and can only be overcome by modifying the flame conditions.

Various modulation frequencies have been employed. Most convenient from the electronic point of view is normal mains frequency, 50 or 60 Hz. In general, any frequency above about 40 Hz is satisfactory, as the normal flame noise component of such frequencies is low. Frequencies below 15 Hz are definitely unsuitable as there is appreciable noise breakthrough even under the best flame conditions. In some instruments, modulation frequencies of 300 Hz or 285 Hz have been used (the latter to avoid mains harmonics) as there is said to be less noise at higher frequencies. In a double beam spectrometer, the beam is modulated and switched by a rotating beam switch mirror, the supply to the source being d.c. or pulsed.

Absorbance conversion

At any given instrumental setting, the photomultiplier output is proportional to the transmission of the flame. A simple linear amplifier with an output feeding a meter thus provides readings on a linear transmission scale. A contiguous absorbance scale can easily be provided but since $A = \log 1/T$ this is much compressed above absorbance readings of 0.5 and virtually unreadable above 1.0. The errors incurred by a linear transmission scale are discussed in more detail on pp. 106 and 107. Many instruments therefore include a circuit for converting the output to be linear in absorbance. Absorbance values can then be read on the linear scale of a meter or a linear recorder.

The absorbance conversion can be achieved in several different ways, but in a single beam spectrometer it is likely to be totally electronic. A logarithmic amplifier can be employed or the logarithmic decay of voltage appearing across a discharging condenser may be utilized. In a double beam instrument, where the two channels have separate linear amplifiers, the servo-driven potentiometer can be wound logarithmically so that the final balance position is a linear function of the absorbance in the sample beam.

Meter and recorder readout

Unlike the liquid cell in a spectrophotometer, the flame is a dynamic entity, and the sample passes through it in a quantized fashion. Source lamps may also introduce their own short term fluctuations and these together may contribute a noise factor over and above the purely electronic one. A meter will, as faithfully as its inertia permits, attempt to follow this noise and will present to the observer anything from a barely perceptible flicker under the best conditions to a considerable irregular waver under the worst, the latter making a sensible reading impossible. The worst conditions referred to will exist when a high amplifier gain or scale expansion is used to measure a trace element in a part of the spectrum for example where emission noise is contributing to the measured signal. An operator, watching the meter needle for some minutes, could not possibly state with confidence its average position.

A certain amount of electronic damping is permissible in order to reduce the noise amplitude, but if the damping is excessive, the meter becomes sluggish and readings are again meaningless. The damping time constant in the meter circuit should therefore not be greater than one second.

The principle advantage of using a chart recorder instead of a meter is that the pen movements – including noise and drift from all sources – are recorded, and a more accurate estimate of an average position is possible. The recorder thus contributes very positively to the accuracy of the measurement. To ensure meaningful traces, the full scale response time should be not less than one second, and the damping time constant between one and five seconds. A logarithmic recorder can be used to convert the output of an instrument linear in transmission to a trace linear in absorbance.

Digital presentation

The modern trend in scientific instruments is to present the results digitally. This avoids errors in scale readings either on the meter or recorder through parallax, mis-interpolation between scale divisions etc. Some instruments are equipped with digital display, and inexpensive digital display units are now available which can be plugged into certain recorder outlets.

A further error-incurring step – that of calculating concentration – still remains. This is not merely a matter of multiplying by a factor (the absorptivity coefficient) because in atomic absorption, Beer's Law, which states that the absorbance of an absorbing species is proportional to its concentration, is strictly true only at low absorbance values. Deviations are usually apparent at absorbances above 0.5, the calibration curve bending towards the concentration axis. Reasons for this are discussed under Calibration (p.79).

Linearization and Concentration Readout. It is therefore convenient to include circuitry in the readout system whereby this curve function can be straightened, so that the absorptivity factor can be applied in order to obtain a direct readout in concentration.

Calibration graph curvature usually tends towards an asymptote of reduced slope and nearly always ultimately to an asymptote of zero slope. Electronic curve straighteners are of two kinds – those which assume that curvature begins at a certain absorbance value and that the curvature is of a defined shape, and those that assume that all graphs are curved (however slightly at low absorbances) and tend towards a horizontal asymptote whose vertical co-ordinate is the stray light level beyond which no further absorption can take place. Neither are entirely correct, but both are capable of extending the linear range to a degree where factorization for concentration is worthwhile. The first type require three controls – slope, onset and curvature – while the second dispenses with an onset control.

Linearization and concentration readout facilities in the form described can be used in conjunction with meter, recorder and digital display. The inherent noise on the input signal would still, of course, be apparent in the readout as flicker on the meter, noise on the recorder trace or uncertainty in the final one or two digits of the display.

Input Signal Integration. In the case of a sophisticated instrument where digital *printout* of the concentration is desired, the noise component is unacceptable because the printer operates at a fixed moment of time and will not necessarily print out the noise- or drift-averaged result. The signal presented to the printer must have been either averaged or integrated over a predetermined period of time.

The averaging process consists of taking ten or one hundred instantaneous readings over perhaps ten seconds during the sample nebulization period. The average of these is then passed through the whole process and printed out as concentration.

Integration continuously sums the absorption signal over a fixed period of of time. This would appear to give the more correct final result, but is more likely to include a 'flash' which could give an anomalously high result likely to be ignored in all other readouts.

Scale expansion

Scale expansion facilities are useful in atomic absorption instruments for accurate work at both high and low concentrations. The reason for the latter is easy to understand. Scale expansion simply involves an electrical expansion of the presented signal by a chosen factor, e.g. 2, 5, 10 or 20. All the signal, including noise, is expanded, and the expanded signal is therefore not more selective than the normal signal. Only the noise originating from the readout device itself is not expanded and this should be negligible with the possible exception of a chart recorder. However, scale expansion considerably facilitates the reading of small scale deflexions, and is particularly valuable when the noise level is 0.5% absorption (0.002 absorbance units) or less.

Real detection limits are often lower than the quoted sensitivity figure, as already defined, by a factor of five or ten, and thus concentrations close to the detection limit are only measurable by scale expansion means.

In single beam spectrometers with transmission readout, scale expansion is simply achieved by increasing the overall gain (amplifier gain, photomultiplier dynode voltage or even slit-width) by the required factor, then backing off the total signal with the zero control until the reading again becomes 100% transmission. Other instruments usually have a separate scale expansion control. This should be completely independent of the amplifier gain and zero control and should simply increase the signal fed to the readout device like the gain control on an independent external chart recorder.

The use of scale expansion in analysis is discussed in more detail in the sections on determination of trace and major quantities p. 95 and p. 111.

REQUIREMENTS IN INSTRUMENTS FOR PRACTICAL ANALYSIS

With a large number of commercial atomic absorption spectrometers to choose from at the present time, the chemist about to embark on the technique may have some difficulty in deciding which features are important as far as his particular requirements are concerned. There is probably no perfect instrument and, if there is, it is extremely expensive. Furthermore, manufacturers' specifications change at frequent intervals. It is not proposed therefore to review instruments individually, but to point out some of the features which are desirable in a good atomic absorption spectrometer for general analytical work, and in Appendix 2 to list the manufacturers by country.

Sampling unit

It should be a simple matter to introduce the sample into the burner, and hence the nebulizer take-up capillary usually extends outside the instrument. This also enables sample presentation devices (p. 55) to be connected thus making the instrument virtually automatic particularly if concentration printout facilities are also available.

The nebulizer, spray chamber, and particularly the burner, though readily accessible for dismantling and cleaning, should be totally enclosed. This prevents draughts from disturbing the flame, and avoids damage both to the operator's eyes from ultraviolet radiation and to his person should a blow back occur. The flame should be viewable through an ultraviolet cut-off filter. The spray chamber should be constructed so that damage to itself is minimal and does not occur to the rest of the instrument in the event of a blow back.

The nebulizer should be constructed so that the sample only comes in contact with inert plastic or an inert metal. The capillary tube is usually made from platinum iridium alloy, and the body of the nebulizer made in tantalum, or coated with an inert plastic. Stainless steel nebulizers are sometimes used, but these are attacked by most strong acids in aerosol form and are soon irreparably corroded even by solutions containing ferric chloride.

Automatic ignition of the flame is not simply a luxury. It is extremely useful, and inspires confidence in a reluctant operator, particularly if interlinked with the gas controls so that the flame can only be ignited under the correct conditions. Flow sensing devices in the gas lines can be made to shut the instrument down in the correct manner should any supply fail. Although automatic control to this extent is very expensive, it ought to be considered seriously if unskilled or junior staff are employed, but it must be proven to be entirely reliable, as a single failure would negate its whole purpose.

Provided the nebulizer and spray chamber function properly, a long tube to the burner itself is no detriment in operation. Indeed it generally improves flame stability. However, an integral spray chamber burner system is preferred for simplicity and cleanliness.

The burner, and particularly the burner jaws, should be massively constructed, to prevent both overheating (which could cause a blow back) and deformation (leading to variations in atomization characteristics). The burner may be pre-aligned in a horizontal sense, but should be capable of adjustment both for height and angle. Burner heads for different gases should be easily interchangeable. A simple push-on fit is best.

Flow controls and meters are essential for both oxidant and fuel gases. It is an advantage to have a third flow control and meter to be used either for auxiliary oxidant or for more sophisticated combustion mixtures such as hydrogen/oxygen/argon etc. which are used in research in atomic absorption and fluorescence and may be more generally required for routine determination of some individual elements.

It is convenient to be able to connect all gases used, e.g. air, nitrous oxide, acetylene and propane or hydrogen, to the instrument and to select the required pair by valves on the instrument.

Radiation source

All lamps of whatever type require some warm up time, and thus it is useful to be able to switch the lamps on independently of the electronics of the instrument.

Turrets containing up to six positions are usually provided for hollow cathode lamps and perhaps vapour discharge lamps. These may rotate or slide to bring the required lamp into position. It is probably better to leave the lamps stationary but to select the one required by a rotating mirror, though this may introduce two further reflecting surfaces in the optical path. This is certainly better for some tin and lead lamps which operate with molten cathodes, and is also more convenient to operate. Some turret systems hold all or some of the lamps not actually in use on a warm-up current which is lower than the operating current. This may save lamp-life marginally, but does not save time as an equilibrium intensity still has to be attained.

The power supplied to the lamp should be stabilized to 0.1 % or better, and the lamp intensity drift should be better than 2% per hour, particularly for a single beam instrument.

Modulation of the lamp intensity may be electronic or mechanical. The frequency should be not less than 50 Hz, and one which is not either mains frequency or a harmonic of mains frequency probably has more advantages than simply a high frequency such as 300 or 400 Hz.

Optics

The merits of single and double beam optics have been discussed. If stability of better than 1% per hour is essential, e.g. in quality control, monitoring etc., double beam ought to be chosen. Double beam instruments also give better precision for work of the highest analytical accuracy.

The monochromator must cover the range from 185 to 860 nm to include the metal resonance lines. The detector may not, for economic reasons, be sensitive over all of this range, but it should be capable of being replaced, with minimum modification to the instrument, by one that is.

If the alkali metals are to be determined a grating monochromator is to be advised (unless a very expensive prism is offered). A prism monochromator is probably a definite advantage if there is to be some emphasis on the determination of metalloids – arsenic, selenium and tellurium – and a few other metals – zinc, cadmium. In this respect the most important parameter is the resolving power of the monochromator – which varies considerably with wavelength for a prism, but is virtually constant for the grating. For most purposes a resolving power of about 0.1 nm is adequate and a dispersion of 3 nm per mm.

Optical accessories

There is little need for a wavelength scanning device in routine analysis, but it is often useful in investigational work e.g. to check interferences from emission lines near the resonance line or molecular emission from the flame.

A useful accessory to the source/optical system of an atomic absorption spectrometer is an ultraviolet continuum source. This enables continuous absorption or scatter to be measured, both of which cause a spurious increase in the observed absorption signal (see page 91). As the spectral bandwidth of the monochromator is large compared with the width of the absorption line (see Fig. 3), an absorption reading taken over this bandwidth using a continuum source contains only a negligible component due to absorption by the resonance line at the same monochromator wavelength setting. The absorbance value thus obtained can be subtracted directly from that given in the normal way with a resonance line. A deuterium lamp is used for this purpose, as this gives a largely continuous spectrum of good intensity at the lower wavelengths where scattering effects become particularly troublesome.

The ability of an atomic absorption instrument to make measurements in emission has been acknowledged for some time, and it is easy to incorporate the necessary facilities in the instrument. It is equally easy to 'design in' the possibility of rearranging the optical system so that fluorescence measurements can be made but perhaps not so easy to modify an instrument that has not been so designed.

Double channel spectrometers

Several double channel spectrometers (as distinct from double beam) are now available commercially. These allow simultaneous measurement at two wavelengths either in absorption or emission. This requires two separate optical systems – primary source, monochromator, detector – and only the flame is common to both. The data-handling sections are usually composite, so that if the channels are designated A and B at least three forms of read-out can be computed.

A and B Simultaneously. This could be, for example, the determination of two elements, with the object of halving the time and/or volume of sample normally required for the analysis. This would be particularly valuable in the solid sampling methods described in the next section when only minute pieces of sample are available. Emission and absorption could be used simultaneously if required.

A – B. The intensity – or absorbance – difference function may be used to obtain background corrections directly, the one channel being tuned to the element line being measured, and the other to a conveniently near point where the background is of similar strength.

Alternatively this mode can be used to correct for spectral interference, either in the flame by emission, or in the source lamp by absorption by tuning the B channel to the wavelength of the interference and balancing the outputs so that A – B = 0 for a sample which does not contain the element being determined.

A/B. This function is used in procedures involving internal standardization (see page 81). Mixed emission/absorption measurements can be made if required.[136]

Automation

In common with many other analytical techniques, atomic absorption has been under the scrutiny of routine analysts, process control chemists and computer experts, with a view to automation. It would appear that atomic absorption techniques as we know them at the present time are not ideally suited to automation to the extent where the operator can happily leave equipment performing analyses unattended for many hours at a time.

Though double beam optics overcome variations in resonance source intensity, the idea, for example, of leaving a nitrous oxide acetylene flame burning unobserved is not attractive, and the problem of automatically selecting a series of sharp resonance lines by a monochromator (if more than one element is to be determined), though not insuperable, has no inexpensive solution.

Possibly, therefore, the manufacturers who simply provide a device for presenting a small number of samples—e.g. 30 or 50—to the atomic absorption spectrometer (preferably with facilities for 'interval wash' with a blank solution or pure solvent) have taken the process of automation as far as practicable in a laboratory instrument.

In considering automation, different analysts may have different processes in mind, and it may be useful to consider the question in relation to each of the various operations that make up an atomic absorption analysis: (i) sample preparation, (ii) serial dilution, (iii) presentation to the instrument, (iv) multi-element analysis, and (v) data handling.

Sample Preparation. The process of preparing a master solution from solid samples is usually the most time-consuming step in an atomic absorption analysis. This has not yet been successfully automated in any automatic chemistry system. It is not inconceivable that weighing out, dissolution, digestion, filtration and making up to volume could all be carried out without human intervention, but the mechanical problems would be considerable. Perhaps the nearest approach is the automation of the Kjeldahl digestion of plant or other biological materials. Here a portion of the prepared solution would be suitable for atomic absorption analysis after serial dilution.

Metals and some other substances in an already comminuted form might also be dissolved in similar equipment.

Serial Dilution. Equipment is available from laboratory supply houses, (e.g. Griffin 'Diluspence', Fison automatic diluters) by means of which predetermined volumes of any number of samples may be accurately diluted with another liquid, e.g. solvent, buffer solution etc. The use of this equipment saves a considerable proportion of the time that is often spent in pipetting and making up to volume.

Presentation to the Instrument. Accessories are available for most atomic absorption spectrometers enabling the operator to load a number of already prepared and diluted samples, and have them presented to the instrument one by one. In the usual versions of this device, the 'dwell' and changeover times are about equal, but can be varied together from 20 to 30 seconds. During the changeover time between samples, a two-way valve allows solvent or blank solution to be aspirated. This 'interval wash' allows the temperature equilibrium of the flame and burner to be maintained and a true base-line is shown throughout the series of analyses on the recorder chart or printout. It also prevents cross contamination between samples. No electrical connexion is required between this device and the main instrument when a strip-chart recorder is used simply to record absorbance output. An interface is required however if the automatic sampler is also to trigger off an integration or averaging sequence within the main instrument.

With this type of accessory and recorder or printer output, the reading sequences, when the operator would otherwise have to sit for long periods in front of his instrument, are carried out automatically. Resetting of the instrument parameters before recirculation of the samples for successive elements must be done manually.

It would seem that this type of device could be fairly readily combined with a diluter, thus enabling the steps described in this and the previous section to be carried out in one accessory. This may also be achieved by using existing automatic analytical systems (e.g. Pye Unicam AC60 Automatic Processing System, Technicon Auto-analyser) though these often provide more facilities than are required. A simple dilution device for atomic absorption analysis was described by Crawford and Greweling.[93] This consisted of a T-junction capillary tube, one arm being connected to the uptake capillary, the second to the sample and the third to the diluent (a stream splitter working in reverse). By varying the diameter and length of the tubes to the sample and diluent, different dilution ratios could be obtained.

Multi-element Analysis. Although instrumental systems have been proposed or devised for the determination of several elements simultaneously, the idea of multi-element analysis by atomic absorption has not found general application, largely because of the limits imposed by the absorption law itself. The concentration range over which an element can be determined with acceptable accuracy in a given solution is comparatively small. It is unlikely therefore that when two or more elements are to be determined simultaneously in the same sample solution their concentrations will all occur in the correct range. Unlike emission techniques, where the sensitivity of individual detectors can be varied, and atomic fluorescence, where sensitivity can be increased with source power, in atomic absorption there is little that can be done instrumentally other than investigating the possibility that a compromised set of conditions may allow all the elements to be measured.

The flame itself may cause most difficulty and this must be chosen in order to atomize all the required elements. It is likely that a hot flame will be necessary with the possibility that some elements may still be incompletely liberated from their compounds while others may be largely ionized. Full use must therefore be made of releasing agents and ionization buffer systems. Sensitivities in the chosen flame can thus be altered relatively only by the use of alternative resonance lines.

If all these difficulties can be satisfactorily solved for a given elemental system, it remains to design an instrument in which the resonance radiation of each element is passed through the flame, and then made to fall on a photodetector having been separated both from all other modulated radiation from the source and from the comparatively very large amount of unmodulated emission from the flame. The latter consideration makes it essential that the dispersing device (monochromator, filter or resonance detector, p. 65) comes between the flame and the photodetector. Unless the incident resonance radiation from all elements is made to pass along the same path through the flame, a separate dispersing device is required for each element.

Walsh (1967) suggested systems whereby coincidence of resonance beams could be achieved.[416] Apart from the simplest one using a multi-element lamp, these usually resolve into the use of two monochromators – one used in reverse to combine beams from individual sources, so that they may pass coincidentally through the flame, followed by the second monochromator used to separate the beams again after the absorption process. Mavrodineanu and Hughes[259] described instruments based on this principle, and also two optical arrangements in which the same grating is used for both functions. Further detectors may be provided if emission measurements are to be made simultaneously.

The use of resonance detectors for each element would much simplify the optics, but, although one commercial double channel instrument using these devices was made available the principle has not become popular.

Data Handling and Computerization. Devices which linearize the calibration relationship and provide a concentration value for display or print-out have already been mentioned. Although sometimes styled 'computers', these are comparatively simple electronic calculators which are set up by hand from data obtained from running calibration standards in order that true concentration readout may be obtained.

The interfacing of atomic absorption spectrometer outputs with true computer facilities opens up more exciting possibilities. Calibration relationships may then be set up by the computer itself and results printed out rapidly in any chosen format. A future step might be the programming of interference effects. One calibration could then be used for a given element in a variety of different matrices, and the matching of standards to samples in which several major components may vary between wide limits would no longer be necessary.

SOME NON-CONVENTIONAL ATOMIZATION SYSTEMS

The nebulizer-spray chamber has been adopted as the best method for introducing a sample into the flame. Its main virtues are stability and, in conjunction with the appropriate gas mixture and burner, very high reproducibility and atomization efficiency. In conventional burners, however, any one atom remains in the light path for a very short period of time. It would seem that sensitivities could be improved if the atomic vapour could be constrained to remain longer in the resonance beam and also if the sample could be introduced without an inefficient nebulization process, or indeed without any pretreatment involving dilution at all.

A number of devices have been described by individuals, and some are available commercially, which take advantage of one or both of these ideas. Without exception they are unsuitable in one way or another for accurate routine quantitative analysis, and some of the more useful are included here under the title 'non-conventional' so that the reader may try them for himself if they appear to provide a solution to a particular problem.

Long tube absorption cells

Although actual gas-flow rates cannot be appreciably decreased to improve the dwell time of atoms in the flame (these are dictated by flame propagation velocities) the combustion products themselves, including free atoms, can be directed along a tube, the axis of which coincides with the optical axis of the spectrometer. The success of a device of this nature depends upon the mean lifetime of free atoms. This, in turn, varies with the element, the temperature and composition of the flame gases.

A simple tube system in which the flame from a total consumption burner is directed in at one end, and the combustion products emerge from the other was described by Fuwa and Vallee.[146] The sensitivity is not proportional to the tube length for most elements as their times of transit exceed the lifetimes of the free atoms. It can, however, be improved to some extent by proper choice of the tube material and by heating the tube with a second burner or an electrical heating jacket. The tube material should be capable of withstanding considerable thermal shock and must not be corroded by the combustion products.

Silica, alumina and ceramic tubes up to nearly 100 cm in length have been used. Such tubes, especially silica, act as light guides because of the reflectivity of the internal surface. Sensitivity may therefore fall off as material from the flame condenses. Condensation also causes memory effects, i.e. the absorption falls off only gradually after the sample has ceased to be aspirated. A further difficulty is that background absorption with tubes of this type is considerable, probably because of the presence of more molecular absorbing species and the greater effect of scattering in a long absorption cell.

A successful long tube device, described first by Rubeska and Stupar[339] consists of an alumina tube 45 cm long, with one end cut at a slight angle so that the flame can enter when the burner is slightly tilted (Fig. 24). The burner may be a total consumption type, or simply the 'emission' burner from a spray chamber system. Best results have been obtained with an air hydrogen flame. It is desirable to be able independently to heat the tube up to about 600° C for some elements.

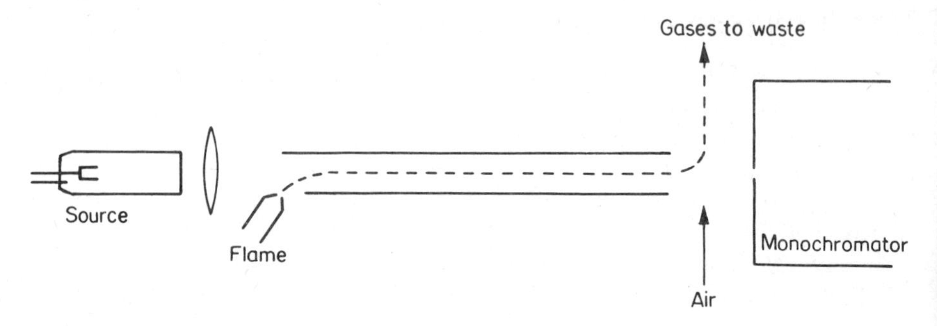

FIG. 24. Long tube absorption cell.

A stream of air must be arranged to blow the combustion products to an exhaust system.

The best sensitivities for all possible elements are achieved when working under fuel-rich conditions. These occur when the burner is so positioned and the flow rates of hydrogen and air adjusted so that the primary and secondary reaction zones are at the entrance and exit to the tube respectively, leaving the interconal zone to occupy the tube. The sensitivity enhancement then improves with increase in dissociation energy of the metal–oxide bond.

Improvements in sensitivity by a factor of approximately 50 over standard flames may be expected for some elements, e.g. silver 0.002, gold 0.01, bismuth 0.025, cadmium 0.001, copper 0.0075, manganese 0.01, lead 0.02, zinc 0.0015, tin 0.04 (all in p.p.m. per 1% absorption) and others. The operation and results given by this system are described in more detail by Rubeska and Moldan.[337, 338] These workers also mention that some elements which form stable oxides give improved sensitivity in the tube, which under the conditions mentioned protects them to a great extent from atmospheric oxygen.

Complete protection from atmospheric oxygen has been obtained by the use of a long absorption tube with separated flames (p. 32). As used by Hingle et al.[171] a long side arm was provided on the flame separator, along which the partly combusted gases were drawn. An optical window was provided at each end, and one tube was surrounded by a furnace which maintains its temperature at 1100° C. Combustion gases were air acetylene with auxiliary hydrogen. Sensitivities using this separated flame system are at least comparable with the other long tube systems, but there is less background, less noise, and less deposition of material in the tube.

Flame adapters

The long tubes described in the preceding section suffer from the disadvantage that they cannot be accommodated in the burner position of a standard atomic absorption spectrometer. A device which uses the existing long path burner both to atomize and maintain the element in the free atomic state, and which can also be made to handle small solid samples directly, is the tube flame adapter. As developed by White[432] this consists of a nickel tube about 10–12 cm long and 1 cm diameter supported in the optical axis with the burner positioned about 2 cm below. A hole 5 mm diameter is made midway along the wall of the tube nearest the burner (see Fig. 25). The flame is lit, heating the tube, and the sample is introduced on a platinum loop or a miniature crucible[123] between the flame and the hole. The

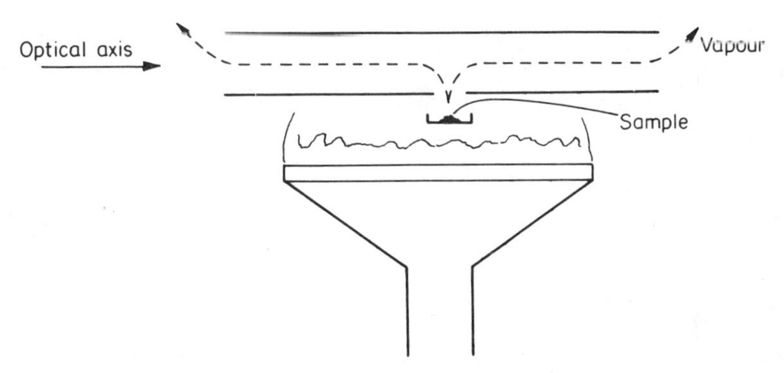

FIG. 25. Tube flame-adapter.

sample becomes partially atomized, and the atoms are retained in the optical path by the tube for considerably longer than with a conventional burner.

Only comparatively cool flames such as air propane may be used with this device, otherwise the tube itself would be melted. It is therefore limited to use with those elements which are atomized at low temperatures, such as lead, cadmium etc.

Very high sensitivities may be obtained for lead – e.g. $< 10\ \mu g/100$ ml in a 20 μl sample – but since reproducibilities are not of the highest order, its use as a quantitative technique has been limited to the screening of lead in blood.

The sampling boat

Somewhat simpler in concept, but based on the principal of evaporating the sample completely and quickly in the flame in order to record a high narrow absorption peak, is the sampling boat.

The boat itself as first described by Kahn et al.[195] is about 5 cm in length and a millimeter or two wide and deep. It is made of tantalum because of the resistance to heat and good conductivity of this metal. A multislot burner is preferred as it heats the boat more uniformly and produces a smooth flow of the combustion gases around it. A simple device is made to enable the boat to be loaded about ten centimetres away from the flame, then pushed close to the flame to dry the sample, and finally placed in the correct position within the flame.

If the sample is aqueous, it must be completely dried before it enters the flame, otherwise some is lost by forceful evaporation. Many samples can nevertheless be prepared as solutions, with standards made in the same way. Solid samples, particularly those containing organic matter, can either be previously wet-ashed, or – more conveniently with the very small samples for which the technique is designed – dry-ashed in the boat itself close to the flame.

Although the air acetylene flame may be used, the sample itself can never become as hot as the flame and the method is therefore limited to elements which are readily atomized such as arsenic, cadmium, indium, lead, mercury, selenium, silver and thallium. Fortunately, these are also the common toxic metals and this method has been used[398] to detect small traces of some of them.

The absorption response curve reaches its peak in 1 to 2 seconds depending on the element, and returns to the baseline after a total time of about 4 seconds. The instrumental electronics and, in the absence of an integrating circuit, the chart recorder, must therefore have a rapid response, and the area under the response curve should be summed or integrated.

Detection limits depend upon the type and volume of sample used. A typical boat, capable of holding 1 ml of solution before it is dried off, could give a detection limit of 0.0001 μg/ml for silver or cadmium – some fifty or more times better than the conventional method.

NON-FLAME ATOMIZATION

Combustion flames, though cheap to produce, stable in operation and, depending on the gas mixture used, able to give a wide range of temperatures, nevertheless have certain serious disadvantages. Chief of these is that the atomic vapour always contains other highly reactive species. It is therefore not possible to predict with any certainty exactly how a given mixture of elements may respond in absorption or indeed how a non-absorbing species may affect or interfere with the elements to be measured. Many attempts have been made to produce the atomic vapour in a completely neutral or unreactive medium, and to introduce the necessary amount of heat energy into the system various electrical methods have been proposed. Almost without exception the initial cost of the electrical equipment required for these methods is comparatively high, as, in order to avoid the effects of selective volatilization, large amounts of energy must be dissipated in a time which is short compared with the total time of measurement.

Non-flame methods fall into two main categories, furnaces and electrically induced plasmas.

The L'vov furnace

As first described by L'vov (1961)[240] a simple graphite tube-furnace, into which the sample was completely vaporized by heating with a d.c. arc, enabled the atomic vapour to be maintained in a highly reducing atmosphere. The tendency for refractory oxides to form was therefore eliminated. High sensitivity was assured as all available sample was atomized into the light beam.

The original furnace consisted of a graphite tube about 5–10 cm long and with an internal diameter of 3 mm. It was lined with tantalum foil and electrically heated to about 2500° C. The sample was supported on a graphite electrode which fitted into a hole halfway along the underside of the tube. A 50 A d.c. arc was struck between the sample electrode and an auxiliary electrode brought up outside the furnace. The sample vapour was prevented from diffusing through the graphite tube by the tantalum lining. The whole device was contained within an argon-filled chamber.

In practice a number of support electrodes, each charged with sample, were also placed in the chamber. The furnace heating current was switched on, and when the furnace had attained working temperature, the support electrodes were inserted consecutively into the opening, and arced for about 4 seconds. The absorption was recorded during that time.

Very small samples, e.g. 0.1 mg, could be analysed as the ultimate detection limits were extremely low, e.g. Sr, 0.02; Ba, 0.1; In, 0.08; Cr, 0.03; Li, 0.05; K, 0.01 – all in nanograms. The length of the graphite tube was chosen so that the volume of sample vapour did not exceed the volume within the furnace tube. The evaporation of the sample had to proceed at a faster rate than diffusion through the ends of the tube, so that there occurred one instant in time when all the vapour was contained within the tube, and the maximum absorbance recording was proportional to the concentration of the element being measured.

An improved method for vaporizing the samples was described by L'vov[241] in 1969. Contact resistance heating of the sample at the point where the support electrode enters the furnace tube results in higher heating efficiency and quicker vaporization.

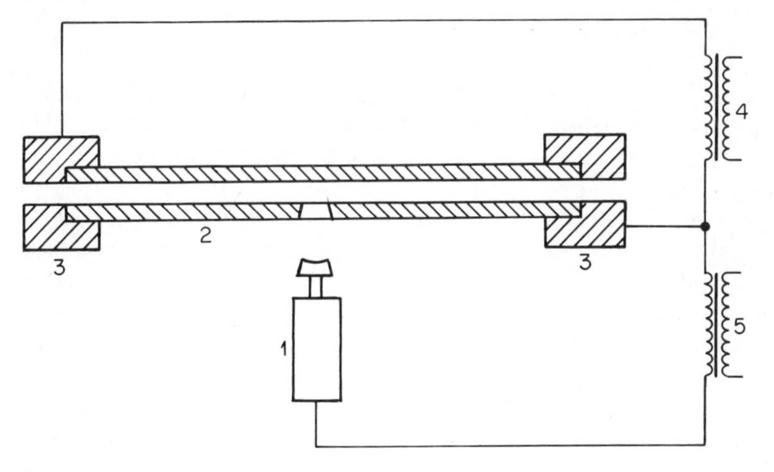

FIG. 26. L'vov furnace.

The system is shown diagrammatically in Fig. 26. The sample is placed on the electrode 1 which can be introduced into the hole in the graphite tube 2. The tube is heated by current from a step-down transformer 4. Connexions are made from the tube and electrode to a second step-down transformer 5, which is switched on only after the electrode has been inserted. In this newer device, the furnace tubes are lined with pyrographite instead of tantalum. This material has low gas permeability, good heat conductivity and oxidation resistance. It therefore minimizes vapour diffusion and promotes even heating. The whole system is again contained in an argon-filled chamber.

Samples may be in solution or powder form. If solutions are being analysed, the electrode heads are treated with polystyrene dissolved in benzene to prevent soaking into the electrode.

The absolute sensitivity values obtained for this method vary between 10^{-14}–10^{-13} g for volatile elements like zinc, magnesium and cadmium to 10^{-11}–10^{-10} g for boron, titanium, etc. L'vov claims that, due to complete evaporation of the sample, many of the interference effects encountered in flame methods do not exist with the furnace, and thus the need for using standards with a composition close to that of the specimen being analysed is eliminated. Precision. however is not as good as for flame methods. Coefficients of variation of 3% would be exceptionally good, and 5–8% are more realistic.

The theory and practice of direct thermal atomization have been discussed in detail by L'vov.[11, 242]

Resistance-heated graphite tube furnace

With the same object in mind, a graphite tube furnace was described by Massmann[254] and later in its commercially available form by Manning and Fernandez.[246] The tube's dimensions are similar to those in the L'vov furnace. It is supported within a water cooled metal container which can be flushed with argon. The tube is heated by passing 400 A at low voltage directly through it producing a temperature up to 3000° C. The heat capacity is low enough for the heating cycle to be repeated for each sample.

Samples are best handled in solution. Small volumes e.g. 10–100 μl are introduced into the tube through a hole in the top by means of a syringe. The sample is dried off before being evaporated. Higher sensitivity can be obtained by introducing three or more 'doses' for one evaporation, drying off each before another is added.

The heating cycle to be used depends upon the type of sample being analysed. With a pure aqueous solution, the current is increased slowly to evaporate the water, then full power is applied to evaporate the sample itself. Biological samples require a slow increase to a somewhat higher temperature in order to char and remove the organic matter – which can be seen as smoke coming from the furnace – before full power is applied. It is normal to record the absorbance during the whole heating cycle so that these phases may be followed. One cycle, after the sample has been introduced, may take between one and two minutes.

The repeatability of this method is extremely good and absolute detection limits found in the author's laboratory, in $g \times 10^{-12}$ are: aluminium 5, arsenic 500, cadmium 0·3, cobalt 10, copper 4, chromium 10, iron 5, lead 5, magnesium 0·01, manganese 3, nickel 2·5, silicon 10, tin 100, zinc 0·02. It is thus readily possible to determine ng/ml concentrations of many elements in 20 or 100 μl samples.

Heated rod ('carbon filament')

The sample may be placed on a carbon rod or filament which also is heated directly by a high current. The filament is placed just below the optical axis so that the vapour is carried upwards and its absorbance recorded.[431] The device is contained within a bell-jar with optical windows. The bell-jar is flushed with inert gas for the measurement, and quickly removed to change the sample.

This system is much simpler to construct than the tube furnaces but is probably somewhat less sensitive because the atomic vapour is not actually contained within the optical path.

Cathodic sputtering

A demountable hollow cathode-like device, into which metallic samples could be introduced, was described by Gatehouse and Walsh.[151] The samples had to be machined to form an open-ended cylinder, and then positioned carefully in the optical beam in the sputtering chamber. This was then sealed, pumped down and filled with argon at the correct pressure. The discharge was then switched on with the sample as the cathode. The absorption of the sputtered atom cloud was measured. This system does not show particularly high sensitivity, and preferential sputtering of certain elements gave rise to some lack of reproducibility.

Plasma torch

Induction coupled plasmas are able to volatilize the most refractory materials as an electronic temperature of something like 16000 K is produced at the centre of the discharge. The temperature decreases along the tail flame and it should in principle be possible to select that part in which conditions are optimum for a given element. Wendt and Fassel[426] showed that a number of highly refractory metals including niobium and tungsten would give sensitivities in the part per million range when solutions containing their salts were nebulized ultrasonically in a stream of argon. The performance was thus at least equivalent to a nitrous oxide acetylene flame. It was also shown that there was no interference of aluminium or phosphorus in the determination of calcium.

While this atom source would seem to have much to commend it – easy sampling, direct sampling of powdered solids, freedom from chemical interferences – the main disadvantages are the very high initial cost, instability ('flicker') in the tail flame and the high degree of ionization of most elements in more stable but hotter parts of the discharge.

REQUIREMENTS IN FLAME EMISSION AND FLUORESCENCE

A flame emission spectrometer consists of an atom source, monochromator and detector, while atomic absorption and fluorescence spectrometers consist of the same parts with the addition of a resonance radiation source shining through the flame and placed respectively on or at right angles to the pre-monochromator optical axis.

It is therefore not surprising that users wish to employ their equipment for two or all three of these techniques. It is also desirable because some elements show their highest sensitivity in emission, some in absorption, and some in fluorescence.

Both 'emission' (by which is understood thermally excited emission) and fluorescence are essentially emission phenomena, the radiation being emitted in all directions. Sensitivity is therefore improved if the optical system is arranged to gather light over the widest possible angle. A monochromator with a large aperture (e.g. f/4) and possibly with light-gathering mirrors placed behind the flame is therefore used. Longer slits can also be used in emission and fluorescence than in absorption in order to allow more wanted energy to reach the detector. Because emission spectra are more complex than absorption spectra, a monochromator with higher resolving power is needed if elements are to be determined in mixtures by emission. Conversely, as fluorescence spectra are simplest of all, measurements in fluorescence can be made adequately with lower resolution.

Emission

The only requirement for flame emission measurements, other than a circular Meker-type of burner head, is that the emission from the flame must now be modulated at the same frequency as the amplifier. This is achieved with a vibrating or rotating chopper between the flame and the detector which is brought into action when the instrument is switched into the 'emission' mode. Light from the hollow cathode lamps used in absorption must of course be prevented from reaching the detector.

Particular developments engendered by atomic absorption have re-stimulated interest in flame emission spectroscopy after a dormant period. Chief of these are the use of the nitrous oxide acetylene flame and of various separated flames. There is thus growing a modern literature on flame emission spectroscopy which features elements that would not have been considered less than ten years ago, but which is beyond the scope of this book.

Fluorescence

Requirements for fluorescence measurements are more exacting. Although the fluorescence intensity depends directly upon the number of free atoms present in the flame and is therefore subject to the same interference effects as atomic absorption, it suffers from one further effect – that of quenching. By 'quenching' is meant the reduction in quantum efficiency caused by the presence of other species in the flame which exert a deactivating effect on the excited atom. Both nitrogen and hydrocarbons give rise to quenching, and therefore in fluorescence it may be necessary to consider combustion gases which do not contain either of these materials. Hydrocarbon gases may be replaced by hydrogen, and air or nitrous oxide by argon oxygen mixtures. Unfortunately hydrogen–oxygen–argon flames do not have the same reducing properties as flames which produce carbon-containing radicals and they are not as effective for the metals which form refractory oxides. The flame gases may therefore have to be chosen more carefully with these two conflicting requirements in mind.

Another quenching effect, known as 'self-quenching' depends upon the relationship between the fluorescence intensity and the atomic absorption coefficient. As derived by Winefordner,[452] the full relationship is somewhat complex, but it was shown that the intensity passes through a maximum value as the number of fluorescing atoms increases. This results in calibration curves which flatten and bend over towards the high concentration values. It is therefore necessary to work in the lower concentration ranges. The critical concentration varies of course from element to element, and even below this value a very wide concentration range (e.g. $1:10^4$ or more) can be handled.

An important condition for sensitivity in atomic fluorescence is that there should be a low level of scattered light – particularly of the same wavelength as the fluorescence. This is again largely achieved by modulating the primary source and the amplifier at the same frequency so that continuous emission from the flame is not recorded. Light scattered by small solid particles is not prevented from being recorded as this will also be of the same wavelength as the primary source. This condition is thus more difficult to fulfil when resonance fluorescence is being measured, but the problem does not arise if stepwise or direct line fluorescence (see page 10) can be utilized, as the fluorescence is then of a different wavelength from the primary radiation and scatter – which is therefore rejected by the monochromator.

From a practical point of view, there are two further advantages of fluorescence over absorption. Since the monochromator only 'sees' the actual fluorescence radiation, and direct radiation from the source is directed away from the monochromator, the primary source does not specifically have to produce sharp line spectra. Therefore more intense sources such as microwave electrodeless or vapour discharge lamps can be used with a proportional increase in sensitivity, the fluorescence intensity being directly proportional to the intensity of the primary radiation.

Taking this one step further, it is clear that fluorescence may equally well be excited by a continuum source, as the free atoms will only absorb at their resonance wavelength. Continuum sources have been successfully used in analytical atomic fluorescence both by Dagnall et al.[104] and by Veillon et al.[412] However, the energy per unit spectral bandwidth for continuum sources – even the high pressure xenon arc – is much less than for a line source over the resonance absorption line width, and so sensitivities are much lower. But a single source can be used for nearly all elements.

DEVICES BASED UPON ATOMIC FLUORESCENCE

Resonance radiation detector

Principle. A resonance detector (sometimes called a resonance monochromator) is a device for detecting the resonance radiation of an element, based upon the fluorescence emitted when the radiation is passed through it.

Resonance detectors were originally based on the design of hollow cathode lamps. A cloud of atoms of the elements to be measured was produced by sputtering from a conventional hollow cathode. Light from a second hollow cathode or

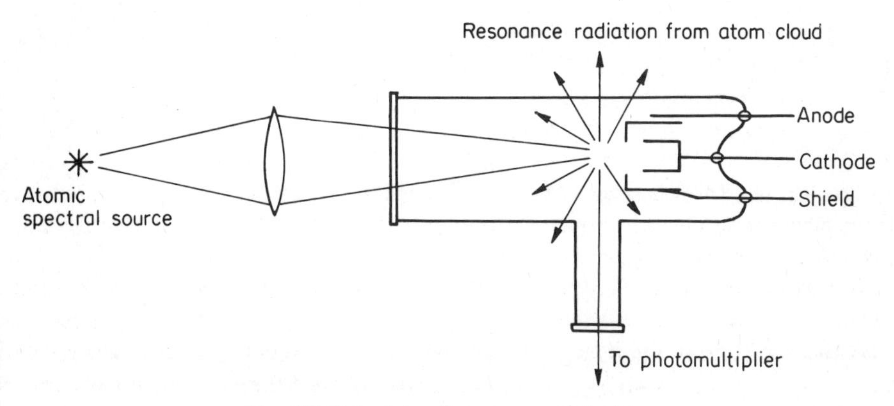

FIG. 27. Principle of resonance detector.

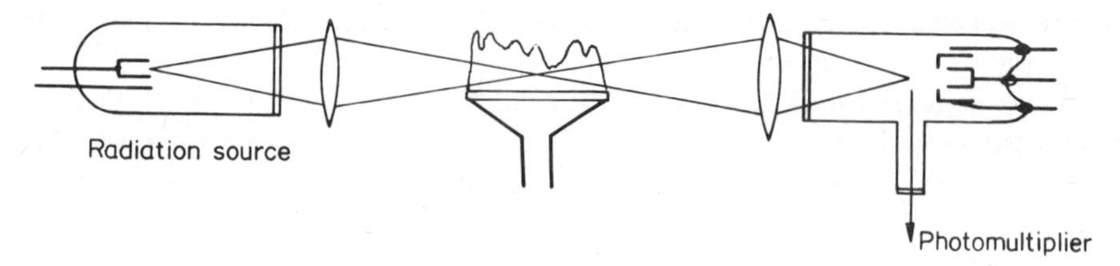

FIG. 28. Resonance spectrometer.

other source passed through the cloud, exciting the atoms to emit fluorescence (Fig. 27). The fluorescence was measured with a photomultiplier. If a sample was aspirated into a flame between the source and the resonance detector (Fig. 28), any of the element present in the sample would cause absorption of the radiation falling on the detector, and there would be a proportional loss of fluorescence intensity.

The most important characteristic of a resonance detector is that it cannot be put out of adjustment by changes in temperature or mechanical shock. The effective resolution of a resonance detector is about 0.001 nm but the optical aperture is wide as compared with a conventional monochromator. Thus the amount of radiation from the flame that can produce noise on the output signal is far less than with a dispersion monochromator.

A disadvantage of the device is that it has a limited life in the same way as a hollow cathode lamp. Also, the conventional type of cathode may need careful shielding to prevent direct passage of light to the detector. These disadvantages are overcome for a very limited number of elements by the thermal type of resonance detector (Fig. 29). Here the atom cloud is produced thermally, the lamp

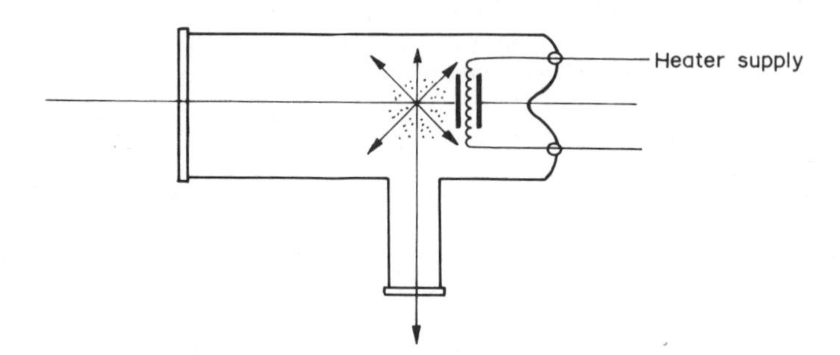

FIG. 29. Thermal detector.

requiring only a few watts to produce the optimum vapour pressure for volatile elements such as calcium, magnesium, copper, zinc and silver.

Uses of the Resonance Detector. Resonance detectors have been used to isolate lines used in the determination of lithium,[69] sodium, potassium, magnesium, calcium, copper, zinc, silver, nickel, lead[416] and other elements. The use of resonance detectors simplifies the optics of multi-element systems, particularly if two elements can be atomized in each detector. Fig. 30 shows one arrangement whereby four elements could be measured at the same time.

The wavelength stability of these detectors has led to their being suggested for portable analysers. A single detector, containing calcium and magnesium, could be furnished with photodetectors placed behind filters to select the wavelengths respectively of these two elements. A dual element source for calcium and

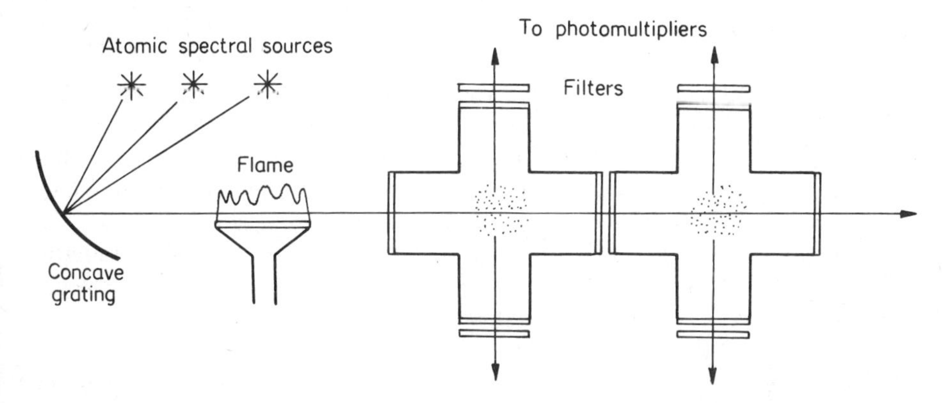

FIG. 30. Multi-element system with resonance detectors.

magnesium would complete a very simple device for determining these two elements.

A device for determining copper, zinc, silver, nickel, cobalt and lead locally in ores overcomes relative sensitivity differences by taking different path lengths through a large circular burner. Sputtering-type detectors are used for all elements except zinc and lead, for which thermal detectors have proved superior. Analytical accuracies of about 5 % are possible.[417]

Simultaneous multi-element fluorescence systems

The simultaneous determination of a number of elements is well known in atomic emission, where arc or spark excitation can be applied easily to solid specimens, but less readily to liquids. The limitations of atomic absorption in multi-element systems have been mentioned, but it appears that fluorescence is uniquely suited to simultaneous multi-element analysis of liquid samples. Some of the reasons for this are given.

Atomic fluorescence spectra are particularly free from spectral interference effects, even with transition and heavy elements.

The atomization and excitation functions are separated (indeed there may be a separate excitation source for each element) so good sensitivities can be attained for each element. In fact the sensitivity of each element is adjusted as required by altering the power of its resonance source.

There are no optical problems in focussing the source radiation into the flame, as, contrary to the requirements in atomic absorption, it can pass in any direction *except* directly into the monochromator.

Further, the concentration ranges of the several elements being determined in one solution are not restricted, either absolutely (provided they are not too high) or relatively, as they would be in atomic absorption by the absorption laws.

A device in which these advantages have been utilized was described by Mitchell and Johansson.[273] This is shown diagrammatically in Fig. 31.

Four element channels were provided, but in principle these could be increased. Light from the four hollow cathode lamps is focussed into the same part of the

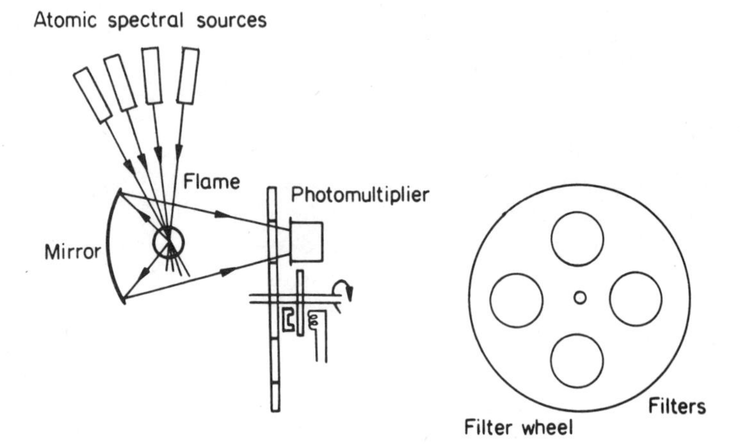

FIG. 31. Simultaneous multi-element fluorescent analysis.

flame, and light from the flame is directed by means of a focussing mirror through a filter wheel on to a photomultiplier. A synchronous motor continuously rotates the filter wheel and an iron pulse-generating wheel at 1 Hz, producing a trigger pulse as each filter comes between the flame and photomultiplier. The wheel

TABLE 4.

Element	nm	Detection limit ppm	Flame
Aluminium	396.1	0.3	N_2O–C_2H_2
Silver	328.0	0.07	Air–C_2H_2
Calcium	422.7	0.003	Air–C_2H_2
Cadmium	228.8	0.03	Air–C_2H_2
Cobalt	240.7	0.06	Air–C_2H_2
Chromium	357.6	0.02	Air–C_2H_2
Copper	324.8	0.04	Air–C_2H_2
Iron	248.3	0.03	Air–C_2H_2
Magnesium	285.2	0.005	Air–C_2H_2
Manganese	279.5	0.003	Air–C_2H_2
Molybdenum	312.6	0.2	Air–C_2H_2
		1.0	N_2O–C_2H_2
		0.1	Argon-separated N_2O–C_2H_2
Nickel	232.0	0.08	Air–C_2H_2
Lead	405.8	0.07	Argon-separated N_2O–C_2H_2
Antimony	217.6	0.1 (in MIBK)	Air–C_2H_2
Selenium	204.0	1.5	Air–C_2H_2
Zinc	213.9	0.07	Air–C_2H_2

holds four interference filters selected to isolate the fluorescence radiation of the four chosen elements.

The trigger pulses are used for gating a 1 kHz modulated supply to the source lamp and the amplifier. The four channels then feed four integrators, the final voltages of which are printed out via a digital voltmeter. The gating pulse sequence is arranged so that the photomultiplier 'sees' the flame emission background in each filter bandpass just before the lamp is switched on, and this is corrected for.

The determination of copper, iron, magnesium and silver in an air hydrogen flame was described, this flame emitting the least molecular background continuum.

A further development of this instrument was described[124, 272] with six channels, a nebulizer burner system for air hydrogen, air acetylene and nitrous oxide acetylene. There were both a.c. and d.c. amplification channels so that either fluorescence or emission could be measured. The detection limits for a signal/noise ratio of 2 are quoted in Table 4.

Chapter 4

Analytical Techniques

INTRODUCTION

Atomic absorption spectrometry is now accepted as a universal method for the determination of the majority of metallic elements and metalloids, in both trace and major concentrations. The form of the original sample is not important provided that it can be brought into either an aqueous or a non-aqueous solution. (It is possible (see pages 58–63) to handle solid samples by special techniques, but in the following pages only quantitative analysis by solution methods will be discussed.) This situation has been brought about by considerable improvements in instrumentation, and also perhaps partly as a result of this, a better understanding among analysts of the types of interference effect that may modify the expected response of a given element.

Atomic absorption methods combine the specificity of other atomic spectral methods with the adaptability of wet methods. High specificity means that elements can be determined in the presence of each other. Separations, which are necessary with almost all other forms of wet analysis, are reduced to a minimum and often avoided altogether, making a typical atomic absorption analytical procedure attractively simple. This fact, combined with the ease of handling a modern atomic absorption spectrometer, makes it possible for routine analyses to be carried out by relatively junior laboratory staff, at a rate hardly foreseen ten years ago.

Usually, separations are required only for one of two reasons – to remove a major cause of interference or to concentrate the elements to be determined should they be present in amounts less than their detection limit. While separation procedures must therefore be quantitative for the elements concerned, they do not necessarily have to be specific as it is possible to determine a number of elements together in one solution. This concept leads to the separation of groups of elements rather than individuals, and indeed to a general philosophy of chemical preparation of samples for atomic absorption in which as many elements as possible are brought together for determination in the final analysis solution. This should always be the aim in method development.

OPERATION OF THE INSTRUMENT

Best results can only be obtained when the measuring instrument itself is maintained in good order and operated correctly. All manufacturers give specific instructions, and usually training, on these matters. However, certain points may be emphasized without in any way duplicating what is laid down in instruction manuals.

It is always desirable to place the instrument beneath a fume extraction hood. This is essential if a nitrous oxide flame or a halogen-containing organic solvent is to be employed or if toxic elements are being determined, in order to remove both toxic vapours and the 3 kW of heat generated by this flame. The fume hood should extend over an area no less than about 30 cm square and the capacity of the extraction fan should be no less than the rate of production of gases in the flame. Its capacity should not be greatly excessive, however, particularly if the flame is not enclosed, as induced turbulence results in a noise component in the output signal. Rarely would it be necessary to remove more than 0.2 m^3 (7 cu. ft) per minute.

Optical alignment

To achieve best performance and sensitivity, the hollow cathode or other source should be positioned, and its image focussed on the monochromator entrance slit, the monochromator and pre-slit optics having been pre-aligned by the manufacturer. With this condition fulfilled, it is wise to check that the burner slot is aligned with the optical path, otherwise a considerable loss of sensitivity may result. Most instruments allow a lateral adjustment of the burner mount for this purpose. The adjustment is best carried out as follows: set up the instrument with no flame and any convenient source line, to read 100% transmission (or zero absorbance) then position a rectangular piece of metal, the 'jig', so that one vertical edge is accurately halfway over the burner slot at the midpoint of the burner length. The jig should now be obscuring half of the incident radiation and the instrument should read 50% transmission or 0.30 absorbance units. If it does not, adjust the lateral position of the burner until these readings are obtained. Then, with the jig placed similarly at each end of the burner slot, rotate the burner about its vertical axis so that the same 50% transmission reading is obtained. Subsequently, when new lamps are to be aligned it should only be necessary to position them so that the maximum output signal is obtained. This is usually done with an adjustment on each individual lamp holder.

Adjustment and care of nebulizer–atomizer systems

The number of free atoms actually produced in the flame 'cell' and therefore the *sensitivity* of the method as already defined depends directly upon the proper functioning of the sample handling system.

Nebulizers may be pre-set or adjustable. If the former, they will have been set in the factory to give the best nebulization efficiency. If adjustable, they can be set by the user to give either the greatest absorbance signal (which should correspond to best nebulization efficiency) or a required uptake rate. The first of these is obviously desirable in the majority of cases and as seen (page 24), it is most likely to occur at a flow rate of 4–4.5 ml/min. Manual adjustment of the nebulizer to achieve maximum sensitivity can be done in one of two ways:

(i) with the flame running and the instrument already recording the absorbance of one of the 'easier' elements, the nebulizer is adjusted until the maximum absorbance signal is recorded.

(ii) Without the flame running but with the air supply connected, a vacuum gauge or simple mercury manometer should be attached to the sample uptake capillary. The capillary is first screwed back until a positive pressure is indicated. (In this condition the nebulizer would 'blow' instead of taking up the sample.) The capillary is then screwed forward until the gauge indicates the first vacuum maximum. It should in fact show only one maximum, unless the capillary is not situated concentrically in the annulus.

Different flow rates may be selected for the following purposes:

a slow flow rate, for use with very small samples, to enable the recording system to reach an equilibrium value before the sample has been used up. The signal would be lower than in proportion to the new flow rate as the nebulization efficiency would also be lower. This could probably be accommodated unless traces near the detection limit were being determined;

a higher flow rate to give a higher absorption signal. This is a more doubtful proposition, as, not only would the increase be less than proportional to the flow rate, but some large droplets may well reach the flame, increasing noise and lowering the flame temperature with consequent loss of stability and reproducibility.

The nebulizer is likely to require little attention provided that it is not attacked or corroded by sample solutions. Modern nebulizers are usually lined with or made from inert materials where the sample comes into contact. They may however become blocked with solid material suspended in the sample, and if not cleaned carefully some damage may result. Solutions filtered through paper – especially 'non-hardened' papers – are a frequent source of blockage, as they contain suspended fibres. Millipore-type filters or centrifugation provide the best means of phase separation, and should be employed where possible. Blockage of the nebulizer capillary itself can be avoided by placing a very short section of similar capillary tube in the sampling end of the uptake tube. Blockage then occurs at this point. The uptake tube is removed from the nebulizer, blown out and then replaced. Blockage of the nebulizer tube itself can usually be removed by passing a suitable length of wire in the sampling end of the capillary and pushing it through to the nose-piece. Wire introduced through the nose-piece may either damage or distort this end of the capillary with consequent alteration of the nebulizer characteristics. Alternatively the nebulizer can be removed from the spray chamber, and, while the air pressure is maintained, a finger is placed over the nose-piece. The air blows back through the capillary and should remove the blockage.

The spray chamber should be kept clean at all times. When purely inorganic solutions are being nebulized this is usually no problem, as the nebulization of pure solvent between and after the samples prevents evaporation of solutions with deposition of solute on the walls and on the flow spoilers.

Biochemical samples, particularly dilutions of blood, serum, proteins etc. tend to cause deposits to build up on the spray chamber in spite of interval washes. This can cause memory effects and lack of precision in the readings. Spray chambers ought therefore to be cleaned carefully after runs of such samples, and certainly at the end of each working day. It is necessary to wash through well between samples which may cause precipitation in the spray chamber (e.g. high chloride followed by high silver solutions) and a wise precaution, in any case, to dismantle and clean the spray chamber and flow spoilers at least once a week. During this weekly clean, too, the burner stem and any associated tubing should be dismantled and cleaned. Although in theory these should never become wetted by the sample mist, in practice some deposition does occur, and if left can start spots of corrosion.

Burner shells should be cleaned weekly, though the jaws may require attention at more frequent intervals, possibly after each run of samples if these contain more than one or two per cent of dissolved solids. Some burner heads can be dismantled for cleaning, and often washing with water is sufficient. Organic or carbon deposits can be removed with a plastic domestic dish scourer and polished if necessary with '000' grade emery paper. Neither acids nor chemical polishes should be used as these start microscopic etch pits which develop into corrosion. The use of metal tools, e.g. screwdrivers, scrapers etc. or even razor blades should not be contemplated as their continued use will soon cause a rough surface and alter the burner characteristics.

The slots of burners can usually be adequately cleaned with a piece of stiff paper (an unwanted visiting card is admirable) if it is inconvenient or impossible to dismantle the burner.

The top surfaces of burners are best polished with very fine emery paper, while deposits formed in the holes of Meker burners can be removed with soft wire.

Effects of organic solvents

Most organic solvents can be treated exactly as water, except that problems sometimes arise when non-miscible aqueous and non-aqueous liquids are brought into contact in the drain tube and syphon. Precipitates or thick emulsions may be formed which prevent the steady flow to the drain of the waste liquid. If this is likely to occur the drain tubes should be emptied and refilled with the solvent to be used in the next run of analyses.

In nebulization systems with plastic tubes, spray chambers etc. it must be established that the organic solvent does not soften or dissolve the plastic. Hardened polythene, propylene and polytetrafluoroethylene seem to be impervious to all likely organic solvents, but P.V.C. is soon softened by methyl isobutyl ketone.

Lighting and maintaining the flames

Some details have already been given about commonly used flames (pages 20 et seq.). When lighting flames it is a good general rule to turn on the oxidant first and to turn it off last. Acetylene burning without premixed oxidant produces smoke and particles of soot, and this condition is much to be avoided, even with an extraction hood. The oxidant should first be turned up approximately to the operating level, then the fuel gas is turned on and the flame lit. The flows are then adjusted to their correct levels, and pure solvent is nebulized until the analysis run begins.

If a flame tends to lift off the burner head either one or both of the gas flow rates are too high. However it is always better to have the flow rates too high than too low before lighting, as in the latter case there is more chance of a blow-back. This particularly applies to gas mixtures with a high burning velocity. The same principle applies when the flame is extinguished. The gases should *never* be turned down together otherwise the critical flame propagation velocity will be reached, and, with the burner plates hot, a blow-back is likely to occur. The operating flow rates should always be such, that if one gas is turned off, or fails, the velocity of the remaining gas through the burner slot is still greater than the flame propagation velocity, and the flame does not burn back through the burner slot.

Instrumental settings and the minimization of noise

Among the instrumental settings to be decided upon in any particular analysis are lamp current, amplifier gain and damping, scale expansion factor, slitwidth, perhaps photomultiplier gain and recorder expansion and damping. These are all chosen in conjunction to provide the most noise-free and responsive signal.

Ideally the signal as recorded on a chart recorder should appear noise-free and sharp as in Fig. 32 (a). Noise from various sources produces less distinct peak values (b). Sluggish response (damping) gives rounded off traces (c).

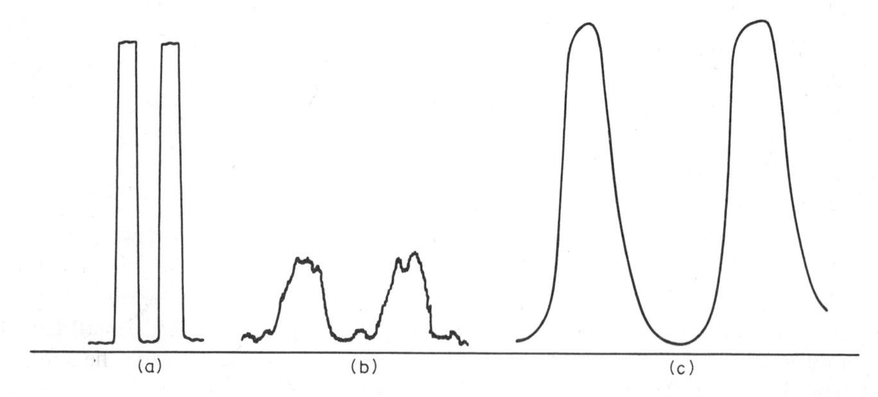

FIG. 32. Types of output trace. (a) Noise free. (b) Less distinct because of noise. (c) Over-damped.

Noise on the signal originates in various ways throughout an atomic absorption system and it is worthwhile to consider these individually to see how they each affect the final signal.

Flame Noise arises from refractive index variations in the region between the hot parts of the flame and the cold surrounding atmosphere and from small variations in effective path length of the flame cell. The flame is a highly dynamic system which cannot be contained within closely defined dimensional limits, and the convection currents induced around it cause such variations. Flame noise always exists. It can only be minimized by good design of the burner ('laminar-flow' burners are clearly the best), the flame compartment and chimney, and by preventing draughts from windows and doors from reaching the flame. A fume extraction system with too fierce a draught can cause turbulence effects within the flame compartment.

Nebulizer ('Concentration') Noise is caused by fluctuations in the amount of the element being measured which actually reaches the flame. Nebulization is a quantized process, and the droplets are produced in a slightly irregular stream. The irregularities are smoothed out during the passage of the mist through the spray chamber. There is thus a minimum effective size of the spray chamber and, as indicated earlier, one which is too small, though giving somewhat higher sensitivity values, will almost certainly produce higher noise.

Both flame noise and nebulizer noise result in short term fluctuations of the absorption occurring in the flame, and hence, by whatever means the absorption signal is recorded or amplified, noise from these sources will remain a constant proportion of it.

Source Lamp Noise consists of long term ('drift') and short term noise. The lamp discharge is a statistical mean of all the micro-scale sputtering discharges which can sometimes even be observed visually as a slight 'flicker'. There is a minimum lamp current below which this effect is so great as to make the discharge unstable. The lamp current must therefore be sufficiently high to minimize this effect, but not so high that broadening of the resonance lines affects the sensitivity.

The intensity of the radiation from the lamp is the subject of all measurements, and any superimposed noise is therefore also amplified and recorded in the final output signal. The only exception to this is lamp noise of a frequency less than the chopping frequency in double beam spectrometers.

Noise Generated in Photomultipliers may be divided into dark noise and shot noise. As the dark current increases with cathode temperature so does the dark noise. While photomultipliers in commercial spectrometers hardly require special cooling facilities, they should not, in a well designed instrument, be subject to undue heating by other parts of the electronic system or by conduction of heat from the flame compartment. Shot noise is caused by irregularities in the electron showers produced by individual photons at the cathode and electrons at subsequent dynodes. The greatest contribution is thus at the cathode, and there may be a small detectable effect at the first dynode. At later dynodes the effect is averaged out, and it is here that the dynode voltage may be varied to produce a higher amplification factor for small incident light signals. As the shot noise is proportional to the square root of the incident intensity, photomultipliers should not be operated close to their threshold intensity values. When operated under ideal conditions, this detector can provide a very high, virtually noisefree, gain factor.

Amplifier Noise in well designed amplifiers with reliable components is entirely Johnson noise. This is so small that it is negligible compared with other sources of noise. This applies to both the gain and scale expansion facilities which, though similar, are separate in instruments reading directly in absorbance. Where a chart recorder is used the only further source of noise is usually the point of contact with the potentiometer or slide-wire. This is readily distinguished by its very irregular character.

In a correctly functioning instrument, the greatest source of noise is often the flame, and this is usually made clear by comparing the character of the signal with and without the flame and sample nebulization.

Among the electronic parts of the instrument, the radiation generator is usually the next greatest source of noise, and the lamp current should therefore be carefully optimized as suggested above. The object of the detector and amplifier is to provide a signal sufficiently large to drive the read-out device: i.e. with no sample, a meter or recorder must indicate full scale deflexion for zero absorbance or 100% transmission; or a double beam system must receive sufficient energy in the reference beam to drive the ratio-computing electronics or null-balance system. The photomultiplier is therefore usually run giving medium gain, and high gain is used only after full amplifier gain has been utilized. In this way it is possible to accommodate a large difference in intensities provided by source lamps of different elements.

Effects of Slitwidth. In the foregoing observations it has been assumed that the width of the monochromator slits has been set to the best value for a particular analysis. The best value is again a compromise. The resolution of the monochromator is improved up to a limiting value (which depends on the remainder of the monochromator optics) as the slits are made narrower. On the other hand more light energy passes through the monochromator and into the detector as the slits are widened. The best value therefore depends on:

(i) the wavelength, if a prism monochromator is in use. The dispersion (in mm per nm) at 220 nm can be about 10 times as great as at 400 nm and twenty times as great as at 500 nm. Hence the slits can be considerably widened at lower wavelengths without loss of resolution but with an increase in energy falling on the detector;

(ii) the nearness to the resonance line of other lines emitted by the source. If non-absorbing lines originating in the spectrum of the element being determined or the lamp fill-gas are included in the spectral bandwidth of the monochromator these will fall on the detector, and however great the absorption of the resonance line, the recorded signal will never fall below their intensity value. In general, therefore, elements with complex emission spectra, e.g. the transition and heavy metals, will require narrower slits than those with simple spectra, such as the alkalis and alkaline earths;

(iii) the intensity of the resonance line. The slit width can be increased if insufficient gain is available from the electrical system to operate the instrument satisfactorily. It may be regarded as a source of noiseless gain available at the cost of resolution and sensitivity. It is noiseless in the sense that it does not introduce a further noise factor. Existing noise is increased only in proportion to the signal.

With a low intensity source, therefore, better stability could be achieved by increasing the slit width, so that a signal of acceptable intensity falls on the photomultiplier. Simply to increase the gain of the photomultiplier operating at a low light level would amplify the higher shot and dark noise proportion of the output. Conversely the sensitivity (working graph slope) is best at the narrowest usable slit width;

(iv) background emission from flame. The level reaching the photodetector is proportional to the square of the slit width and, as has already been noted, increases shot noise and the likelihood of saturating the detector.

Electrical Damping reduces the noise shown by a meter or on a recorder trace. These readout devices themselves provide some inertia in the system but further variable damping is usually provided. The degree of damping is usually defined by the time constant of the noise-filtering circuits. Typically this may be 0.2–2 seconds with a meter, and 1–4 seconds with a recorder. The time required to make a reading is proportional to this time. If the time constant is too long, then at best, very rounded off traces (see Fig. 32 c) or irregularly drifting traces or readings may be obtained. If it is too short an excessive noise amplitude will be apparent. However in cases of severe noise it is often easier to estimate a mean visually when the noise is reduced to say $\pm 3\%$ than it is to rely on the settling-down of an over-damped meter or recorder.

In setting up an instrument it is thus best to select the least damping initially so that rapid responses are obtained. It can then be increased somewhat if necessary to make the final readings or recordings.

Integrating instruments

The noise effects which have been described are observable only on instruments giving continuous readout. In the integrating or averaging systems employed to facilitate digital display or printout, the effects, though nevertheless present, may not be so apparent. The noise is manifest in variations in the averaged readings observed with the shortest integration time. This may be 0.2 or 0.4 seconds and thus gives virtually as much information as a recorder though in a form which may not be quite as readily appreciated. Conditions are adjusted as before until this variation has been minimized. When actual analyses are being performed the time constant is increased to the working value which may be 10 or 20 seconds or thereabouts.

CALIBRATION

Standard working curves

The relationship between the absorption indicated by the instrument and the concentration of the element which produces it is established in the calibration procedure. The theoretically linear relationship between the amount C of the absorbing species in the light path and the absorbance A, known as Beer's Law, is followed within certain limits in atomic absorption spectrometry. After an almost linear portion the calibration curve usually bends towards the concentration axis to a greater or less extent.

If T is the transmission of the flame and I_0 and I the intensity of the resonance radiation entering and leaving it, then by definition

$$A = \log 1/T = \log I_0/I \qquad (1)$$

The output of the detector, being proportional to the energy falling on it, is proportional to T, and hence the conversion from T to A must be made, by calculation, by a non-linear scale on the instrument meter or electronically in instruments provided with 'linear absorbance readout'.

In instruments giving linear transmission, the meter deflexion $d \propto I$, so

$$A = \log \frac{d_0}{d} \qquad (2)$$

The value of d with no absorbing species is usually set to 100, hence the absorbance is 2 minus log of percentage transmission. It is sometimes inconvenient to set zero absorption to 100 on the meter or recorder, particularly when 'blank' absorptions may have to be observed or corrected for, so, provided that the instrument is adjusted so that zero light gives zero deflexion, the zero absorption value can be set with the gain control to any convenient deflexion d_0 and absorbance calculated from equation (2).

Instruments giving linear absorbance readout are automatically adjusted for zero light on a transmission scale, as zero energy corresponds to ∞ A. Zero absorption would then normally be adjusted by amplifier gain controls to $0A$, though, as absorbances are directly subtractive, it could be set if desired to any other convenient value on the scale.

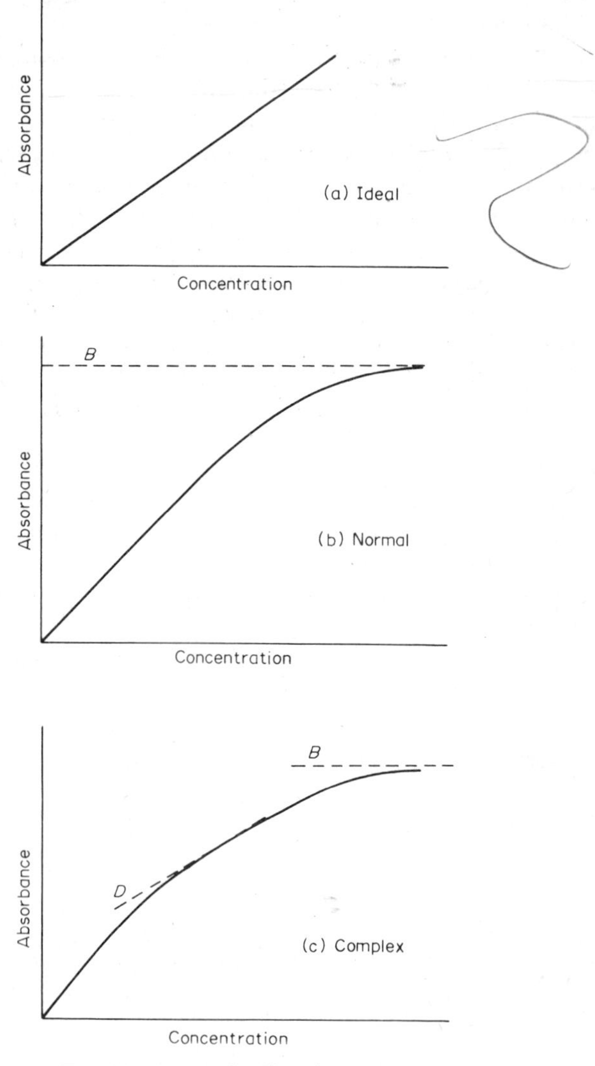

FIG. 33. Types of calibration curve.

In practice several working standards should be prepared to cover the best absorbance range, say 0, 0.2, 0.4 and 0.6 *A* units. If the concentration range corresponds to much higher absorbances, the sensitivity must be reduced (see Chapter 4. Determination of Major Components) or if it corresponds to much lower absorbances other measures (Chapter 4, Trace Analysis) may be necessary.

Non-linearity of Calibration Curves. The calibration curve of concentration against absorbance consists of an almost linear portion near the origin, followed by a section which bends to a greater or less extent towards the concentration axis. By convention, concentration (the independent variable) is the abscissa. The amount of bend, and the point of onset of bend depend on the amount of unabsorbable radiation reaching the detector, and on the presence of less sensitive lines within the monochromator bandpass. Three possible calibration curves are shown in Fig. 33. In (*a*) the ideal (almost impossible) case where all light reaching the detector is absorbed by the element being determined to the same extent; (b) the normal curve, where *B* represents the residual unabsorbed light level, and the curve is asymptotic to this value. Unabsorbable light may be due to non-absorbing lines from either the source cathode material or the fill-gas which pass within the spectral bandwidth of the monochromator. It also occurs if the ratio of the half-widths of the emission line w_e and the absorption line w_a approaches or becomes greater than unity. Rubeška and Svoboda[340] showed that the calibration graph can only be linear if w_e/w_a is less than 1/5. When $1/5 < w_e/w_a < 1/1$ the graph is slightly curved, but if $w_e/w_a > 1/1$ the initial slope (which, as will be seen in the section on Trace Analysis p. 92, is a measure of the analytical sensitivity) starts to be decreased. Stray light of other wavelengths, too, caused by bad monochromator design or light of the correct wavelength which has not passed through the area of maximum atom concentration in the flame has the same effect. In Fig. 33 (c)

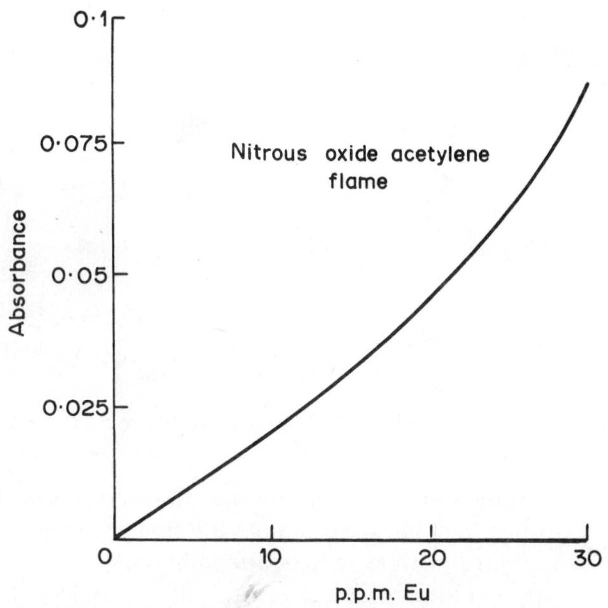

FIG. 34. Upward-curving calibration for europium.

is shown a further, more complex case where a line of lower absorbance occurs within the bandpass, and the calibration curve first becomes asymptotic to *D* its sensitivity and then again to *B* the unabsorbed light value.

Many elements give linear calibrations up to about 0.5 *A* provided the instrument is set up and operated correctly and most of the remainder give only slight curvature to 1 *A*. A notable common exception is iron, which, because of the occurrence of the main resonance line in a densely populated spectrum region, invariably shows curvature which depends on the bandpass and available resolution.

Occasionally, a calibration graph showing distinct curvature away from the concentration axis may be encountered. A typical example is europium in a nitrous oxide acetylene flame (see Fig. 34), or sodium or potassium in the air acetylene flame. The cause is ionization. At very low concentrations a higher proportion of the element is ionized. This is clear from the ionization curves for 10^{-6} and 10^{-8} atmospheres partial pressure in Figs. 6 and 7, if percentage ionization at a given temperature is read off. The effect is noticeable at temperatures above the optimum for a given element. Correction, as will be seen when interferences are discussed, is by addition of an ionization buffer.

Method of standard additions

Sometimes it is not possible to overcome interference effects by matching

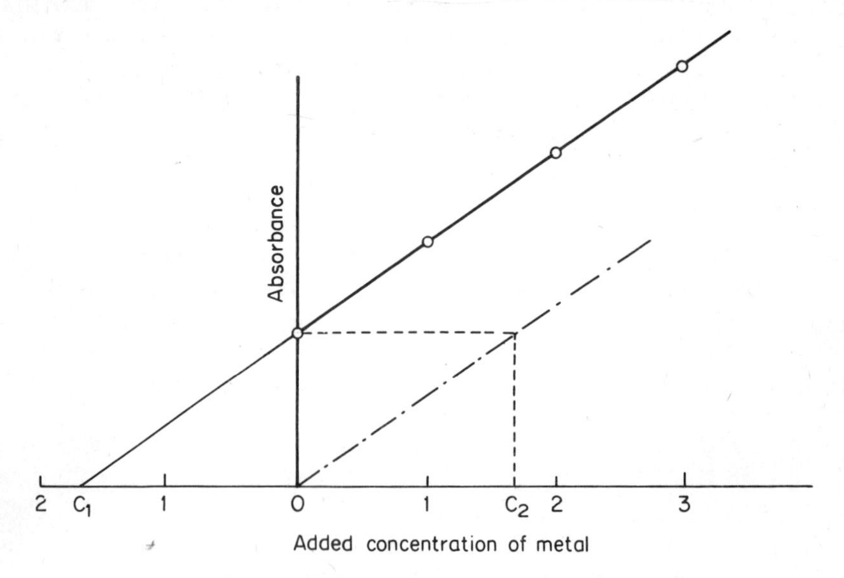

FIG. 35. Method of standard additions.

standards with the samples, perhaps because the full composition of the samples is not known. Provided that sufficient sample solution is available, the analysis can then be done using the method of standard additions.

The sample solution is divided into a number of aliquots – at least three are necessary. To all but one of these are added known increasing amounts of the metal to be determined and the samples then made up to the same concentration. The solutions are nebulized and the absorbance readings plotted against the added concentration of the metal (Fig. 35). The graph obtained is the heavy line, and the absorbance of the unmodified sample is due to the element concentration sought (plus background absorbance and scatter). Ignoring the latter, the *true* calibration curve would be parallel to the one obtained, but passing through the origin. The required concentration may thus be read off at C_2. In fact if the graph drawn is produced backwards, its intercept C_1 on the C axis is, by similar triangles, equal to C_2, and this is where it is most conveniently read off.

The accuracy of an extrapolation method such as this is never as good as an interpolation method but sometimes it is the only one possible. The value found may well include background and scatter, and to improve accuracy this must be determined separately by measuring the absorbance value at a close non-absorbing line as described in more detail on page 91. This value is then subtracted from all the other readings prior to plotting the graph in Fig. 35. It should also be checked if possible that the element being measured gives the same response in the sample as in the additions, particularly if it is likely to be complexed or protein-bound.

Internal standardization

The concept of measuring the ratio of the signals from the analysis line and an internal standard line is well known in emission spectroscopy, where the internal standard is either the major element (as iron in steel) or an added element of known concentration. In atomic absorption, the method can be applied only when a double- or multi-channel instrument is in use, i.e. one which can measure simultaneously at more than one wavelength. Such instruments, though not common, are commercially available.

Internal standardization in atomic absorption was used by Butler and Strasheim[81] in an attempt to correct, at least partially, for variations in nebulizer and flame performance. Such variations are not corrected for by double beam spectrometers. When determining silver and copper in gold using a converted spectrograph, Butler and Strasheim found that the use of a non-resonant gold line as internal standard gave better results than a resonant gold line, and that to plot the *ratio* of the intensities of absorbed copper or silver and non-resonant gold lines gave higher precision than plotting simply copper or silver absorbance.

Feldman[137] has described internal standardization techniques with a double channel spectrometer using absorption and emission techniques separately and in combination. When absorption is used in both channels, light from two hollow cathode sources is made to pass through the same flame. Light of the analysis and internal standard wavelengths is separated by means of two monochromators. The stability of the ratio of output signals was always superior to the analysis signal alone, and reproducibility of analyses as expressed by coefficients of variation was better, usually by a factor of 2–3.

When the internal element is selected, the major criterion is the relative flame sensitivity which should be matched as far as possible for all parameters affecting the elements' absorbance in the flame. An unmatched internal standard element will correct simply for dilution, flow and nebulization changes. Useful analyte –

internal standard pairs recommended by the same author are Ca/Sr; Al/Cr; Fe/Au or Mn; Mn/Cd; Cu/Cd, Zn or Mn; Cd/Mn; Zn/Mn or Cd; Pb/Zn; Si/V or Cr; V/Cr; Ni/Cd; Cr/Mn; Mg/Cd; Co/Cd; N or K/Li; Au/Mn; Mo/Sn.

Preparation of standard and reagent solutions

Standard Solutions. The general technique is to prepare relatively concentrated stock solutions for each element from which working standards are prepared by serial dilution. A stock should be a solution of a simple salt of one metal. Reagents used in preparation of stocks need not be excessively pure as they are to be used at considerable dilutions. Analytical reagent grade is quite sufficient.

For the sake of convenience it is worthwhile to weigh out the exact quantity of the salt to give a stock solution of 1000 p.p.m. of the metal ion. The pure metals themselves are used where possible, particularly if their salts are not stable enough on the shelf to be considered as primary solution standards. The metal should be in the form of foil, sheet, rod, wire or ribbon, and the surface should be de-oxidized either by dipping in acid or by abrasion. Powdered metals and sponges should be avoided as they may contain significant quantities of oxide.

Suitable reagents and volumes for most common metals are given in Table 5. Most of the stock solutions prepared in this way may be stored in polythene bottles for six months or more (gold is a particular exception). Very dilute standards, 1 p.p.m. and less should not be used for more than one or two days.

TABLE 5. Preparation of standard solutions. The following are to be made up to 1 litre with water.

Element	Reagent	Weight, g	Dissolution	Concentration*
Aluminium	Metal, foil	1.000	25 ml conc. HCl + a few drops conc. HNO_3	1000
Aluminium	Metal, foil	0.2647	10 ml conc. HCl + a few drops conc. HNO_3	500 (Al_2O_3)
Antimony	Metal, shot	1.000	10 ml conc. HCl + a few drops conc. HNO_3 + 10 g tartaric acid	1000
Arsenic	As_2O_3	1.320	50 ml conc. HCl	1000
Arsenic	As_2O_3	1.320	20 ml 5% w/v NaOH	1000
Arsenic	As_2O_3	1.320	50 ml 0.880 NH_4OH	1000
Barium	$BaCO_3$	1.438	20 ml M-HCl	1000
Beryllium	$BeSO_4$	19.639	Water	1000
Bismuth	Metal	1.000	50 ml conc. HNO_3	1000
Boron	Boric acid	5.714	Water	1000
Cadmium	CdO	1.142	20 ml 5M-HCl	1000
Calcium	$CaCO_3$	2.497	100 ml M-HCl	1000
Calcium	$CaCO_3$	1.785	25 ml 2M-HCl	1000 (CaO)
Chromium	Metal	1.000	50 ml conc. HCl	1000
Cobalt	Metal	1.000	50 ml 6M-HNO_3	1000
Copper	Metal	1.000	50 ml 5M-HNO_3	1000
Germanium	Metal chips	1.000	20 ml 1:1 HCl:HNO_3	1000

TABLE 5. cont.

Element	Reagent	Weight, g	Dissolution	Concentration*
Gold	Metal	1.000	15 ml conc. HCl + 5 ml conc. HNO_3	1000
Gold	Ammonium chloroaurate	1.811	Water	1000
Indium	Metal	1.000	10 ml conc. HCl + 5 ml HNO_3	1000
Iron	Metal	1.000	20 ml 5M-HCl + 5 ml conc. HNO_3	1000
Iron	Metal	0.6990	ditto	1000 (Fe_2O_3)
Lead	Metal	1.000	50 ml 2M-HNO_3	1000
Lithium	$Li_2SO_4 . H_2O$	9.215	Water	1000
Magnesium	Metal ribbon	1.000	50 ml 5M-HCl	1000
Magnesium	Metal ribbon	0.6040	20 ml 5M-HCl	1000 (MgO)
Manganese	Metal	1.000	50 ml conc. HCl	1000
Manganese	Metal	0.7744	50 ml conc. HCl	1000 (MnO)
Mercury	Metal	1.000	20 ml 5M-HNO_3	1000
Molybdenum	$(NH_4)_6Mo_7O_{24} . 4H_2O$	1.829	Water	1000
Nickel	Metal	1.000	50 ml 5M-HNO_3	1000
Niobium	Metal powder	1.000	10 ml HF + 5 ml conc. HNO_3	1000
Palladium	$(NH_4)_2PdCl_4$	2.668	Water	1000
Platinum	$(NH_4)_2PtCl_6$	2.275	Water	1000
Potassium	KCl (dry)	0.7455	Water	10 mequiv./l
Potassium	KCl (dry)	1.905	Water	1000
Potassium	KCl (dry)	0.1584	Water	100 (K_2O)
Rhodium	$(NH_4)_3RhCl_6 . 1\frac{1}{2} H_2O$	3.858	Water	1000
Ruthenium	$[(OH)(NO)Ru(NH_3)_4]Cl_2$	2.839	Water	1000
Selenium	Elemental	1.000	15 ml conc. HCl + 5 ml conc. HNO_3	1000
Silicon	$Na_2SiO_35H_2O$	c.7.6	Water	1000†
Silicon	$Na_2SiO_35H_2O$	c.8.82	Water	2500 (SiO_2)†
Silver	$AgNO_3$	1.575	Water	1000
Sodium	NaCl (dry)	2.542	Water	1000
Sodium	NaCl (dry)	0.5845	Water	10 mequiv./l
Strontium	$SrCO_3$	1.684	20 ml M-HCl	1000
Tantalum	$TaCl_5$	1.9798	Water	1000
Tellurium	Metal	1.000	15 ml conc. HCl + 5 ml conc. HNO_3	1000
Thallium	Tl_2SO_4	1.235	Water	1000
Titanium	Pot. tit. oxalate	7.3939	Water	1000
Tin	Metal	1.000	200 ml conc. HCl + 5 ml conc. HNO_3	1000
Vanadium	Ammonium vanadate	2.2960	20 ml 100 vol. hydrogen peroxide	1000
Zinc	Metal	1.000	50 ml 5M-HCl	1000
Zinc	Metal	0.804	20 ml 5M-HCl	250 (ZnO)

* Concentration as metal p.p.m. unless otherwise stated.
† Must be standardized gravimetrically.

Reagent Solutions are those which are added to standards and samples to overcome interferences, or simply to the standards in order to match the samples. They are invariably used in concentrated form, and therefore they must be of a high degree of purity, particularly with respect to the element being determined.

Lanthanum chloride is used extensively as a spectroscopic buffer as will be seen when interferences are discussed. Ordinary grade lanthanum salts are usually heavily contaminated with calcium and magnesium, the very elements which would benefit most by the use of lanthanum, and the amount of these elements added in the buffer may well be more than those being determined. Special atomic absorption grade lanthanum chloride and lanthanum oxide is now available from some of the major reagent manufacturers, which is especially low in calcium and magnesium (e.g. 0.2 and 0.1 p.p.m.). Additionally, a number of the alkali metal salts used as ionization buffers, strontium salts and EDTA (releasing agents) mineral acids and other dissolution reagents and extraction solvents are also available in 'atomic absorption quality'.

When the composition of a standard has to be matched with that of a sample, the matrix element to be added must be in a high state of purity, as there is no dilution of its own impurity elements compared with the sample. Only the purest forms available, e.g. 'Specpure', may then be good enough. Although 'blank' corrections can be made, the accuracy of the method falls off severely as the 'blank' value approaches the concentration being measured.

Non-aqueous Standards. When samples are dissolved in organic liquids, calibration can often be effected by dissolving suitable organo-metallic compounds in the same solvent. The choice of solvent may well be in the hands of the analyst, particularly if its purpose is simply that of dilution, and, in this case, methyl isobutyl ketone or white spirit are often preferred, as their physical properties approach those of water.

The solvent has a profound effect on both the physical and chemical properties of the flame and thus influences the precision and sensitivity of the analysis. Many common organic solvents are unsuitable for spraying into the flames produced by the usual nebulizer/spray chamber system as they are incompletely combusted and produce smoky yellow flames. The most suitable solvents are C_6 or C_7 aliphatic esters or ketones and C_{10} alkanes. Aromatic compounds and halides are usually unsatisfactory for the reason given, and the simpler solvents (methanol, ethanol, acetone, diethyl ether, lower alkanes etc.) because of their tendency to vaporize in the nebulizer causing an erratic response.

A number of organo-metallic compounds are now available for the preparation of standards. Metal naphthenates have been used extensively in petroleum oil and paint analysis as these dissolve readily in white spirit. These compounds are available from Messrs. Burt & Harvey Ltd. and a further range of compounds from Eastman Organic Chemicals Ltd. Suitable compounds are given in the section on individual elements (Appendix 1).

SAMPLE PREPARATION

For the purposes of a general résumé of methods of preparation of materials for analysis by atomic absorption, samples may be divided into those which are already in a solution or liquid state, solid samples which contain a high proportion of organic matter and entirely inorganic solids such as metals, rocks etc.

Liquid samples

Aqueous Solutions can sometimes be nebulized without any prior treatment. Among these are domestic, natural and boiler waters, wines, beers and perhaps urine. A prior knowledge of the approximate concentration of the elements being determined is useful as it enables the operator to decide whether or not some dilution is desirable. If the element is in high concentration (e.g. more than 200 times the analytical sensitivity value) the solution should be diluted to bring it within the best concentration range for measurement (20–200 times the sensitivity). If the concentration is very low, scale expansion or some chemical pretreatment may be required.

It may also be necessary to add a spectroscopic buffer to aqueous samples in order to suppress an interference effect. Such buffers are used in concentrated form in order to avoid undue dilution of samples where trace elements are to be determined.

Non-aqueous Solutions can also sometimes be run directly but this depends on whether or not the viscosity is similar to that of water – for which most nebulizers are designed. White spirit and methyl isobutyl ketone fulfil this condition, and hence are recommended diluents for many organic liquids. Mineral oils and petroleum spirit, paints and drying oils, vegetable oils and organic liquids are some types of material that require minimal pretreatment. Standards can be made up in the pure basic solvent.

Solid samples

Organic. Many types of samples consisting of trace or minor elements in a largely organic matrix are best brought into aqueous solution after wet or dry oxidation of the organic material. Such samples include plant material, fertilizers, soils, food and feeding stuffs, biochemical specimens etc. The dry oxidation procedure offers the advantage that the residue may be taken up in almost any desired acid medium, but it cannot be used if volatile elements such as mercury, arsenic, lead, antimony and molybdenum are to be determined. Both oxidation procedures enable the trace elements to be concentrated and standards are made up in the same acid medium as the sample itself finally appears.

Organic chemicals and pharmaceutical products may also have to be treated by the above procedure but it is possible also that they may dissolve either in water or an organic solvent, in which case they can probably be presented to the instrument in this form.

A further successful procedure for organic-based solid samples – or liquids – is acid extraction. The sample usually has to be shaken or stood overnight in contact with hydrochloric or nitric acid. Quantitative extraction has been proved, e.g. for nickel in fats, and a number of trace elements in plant material.

Inorganic. Recognized methods for the dissolution of metals, slags, ores, minerals, rocks, cements and other inorganic materials and products can be used, the analyst attempting as far as possible to bring all the elements to be determined together in one solution.

Where final solutions contain more than about 0.5% of dissolved material, the standards should also contain the major constituents, even where no chemical interference is expected, in order to match the viscosity and surface tension and avoid 'matrix' effects.

INTERFERENCE EFFECTS

Because of the simplicity of absorption spectra, there are very few known examples of actual spectral interference either in atomic absorption, other than monochromator bandpass effects, or in atomic fluorescence.

Interferences which actually influence the proportion of atoms in the flame available to absorb the resonance radiation arise largely from chemical effects which originate either in the flame itself or in the sample solution. Such interferences are caused by the formation of stable compounds or by ionization.

In addition, minor interferences are caused by various physical phenomena including incomplete volatilization of the solid particles of samples formed in the flame, variations in the physical properties of the solutions (matrix effects) and scatter and background absorption.

Any effect which influences the number of free atoms in the flame, be it chemical or physical, affects the results obtained in emission, absorption or fluorescence equally. Absorption measurements are free from those interferences which arise from interactions between excited atoms, e.g., the transfer of energy by collisions of the second kind which affect emission and the similar quenching effects which have been noted in fluorescence.

Chemical interferences

It has been the practice among atomic absorption analysts to differentiate between 'enhancing' and 'depressive' interferences. In all cases however the presence of substances which produce an improved sensitivity for a given element actually do so by reducing an existing depressive interference. It would appear, therefore, to be more logical to look upon any effect which causes an element to give less than its maximum possible absorption as a depressive interference. The maximum possible absorbance would be given when an element is 100% atomized, but this condition, though readily defined, is unlikely to be attained in practice because of the limitations imposed by the atomization and ionization curves (see page 14).

Stable Compound Formation is the best known of the interferences. It is responsible for many of the depressive effects reported over the years in both emission and absorption flame photometry. It arises because compounds or radicals containing the element being measured are not broken down into individual atoms at the temperature of the flame being used. Stable compounds may even be formed in the flame. This condition exists below the atomization curves of Figs. 6 and 7.

Typical examples are the lowering of alkaline earth metal absorbances in the presence of aluminate, silicate, phosphate and some other oxy-anions, the low sensitivity of metals which form very stable oxides (such as aluminium, vanadium, boron etc.) and the depression of calcium in presence of protein. All these occur in the air acetylene flame.

The effect of oxyacids on alkaline earth metals is normally appreciable only when the anion is in a chemical excess over the metal. This suggests that an equilibrium is being set up between the species M–O, or M–Cl if the materials are dissolved in chloride medium, and M–O–X. In the presence of excess –O–X, the equilibrium of the first part of equation 1 will tend to move to the left preventing formation of free M. High absorbance readings are obtained at higher temperatures, and also with hydrocarbon fuel gases as compared with hydrogen. There would therefore appear to be a further mass action effect which, according to the conditions in the flame and concentrations of reacting species in the flame plasma, favours the production of either the free atoms or the stable associated M–O species. Increased temperature results in the formation of free atoms of the wanted metal M according to the second stage of reaction 1.

$$\text{M–O–X} \underset{\substack{\text{excess}\\\text{oxyacid}}}{\overset{\text{heat}}{\rightleftharpoons}} \text{M–O} \underset{\substack{\text{low flame}\\\text{temp}}}{\overset{\text{heat}}{\rightleftharpoons}} \text{M} + \text{O} \overset{\text{C}}{\rightleftharpoons} \text{M} + \text{CO} \qquad (1)$$

Lower temperatures and/or excess of the oxyacid thus favour the persistence of the stable oxy-salt.

The action of a releasing agent is to influence the chemical equilibrium in the desired direction:

$$\text{M–O–X} + \text{R} \overset{\text{excess R}}{\rightleftharpoons} \text{R–O–X} + \text{M} \qquad (2)$$

If R is a metal which forms a similarly stable compound with the oxyacid, and is present in excess, then mass action dictates that reaction 2 must proceed to the right, producing a higher proportion of free atoms.

Good releasing agents are therefore those metals which themselves form stable oxysalts, and for this reason strontium and lanthanum have been most used.

From the second and third stages of reaction 1 it is clear why metals forming refractory oxides give low sensitivity at low flame temperatures. An increase in temperature, such as is achieved when nitrous oxide is used as oxidant instead of air, will itself tend to improve the situation. It has already been seen (p. 21) that the presence of free carbon or carbon-containing radicals up to the critical C/O ratio removes the excess of oxygen as carbon monoxide pushing the equilibrium further to the right. In this way carbon itself behaves as a releasing agent. The practical results, that hydrocarbon based flames give better sensitivity for refractory metals, might thus be predicted.

The effect of metal–oxygen bond energy on the sensitivity, in the nitrous oxide acetylene flame, of a number of metals showing high bond energy was investigated by Sastri et al.[345,346] Solutions of simple salts and oxysalts were compared with solutions containing metallocenes and fluor-complexes where the metal is not oxygen-bonded. In the latter cases the sensitivity of the metal was always higher. An irregular correlation was then shown to exist between the increase in sensitivity and the metal–oxygen bond energy. A further type of releasing action is thus the prevention of the formation of metal oxygen bonds in solution by suitable complexation. The absorbances of titanium and zirconium are improved for example in the presence of hydrofluoric acid.

Another example of a stable compound formed in solution is the calcium–protein complex, which, present in blood serum, appears to persist in the air acetylene flame. Cooke and Price[91] showed that the calcium absorbance is not a direct function of the dilution factor, even when lanthanum chloride is added as the releasing agent. When EDTA is used instead, the sample can be used at any dilution and the correct concentration of calcium calculated.

This is explained as the calcium EDTA complex is stronger than the calcium–protein complex in solution, but the latter is more readily dissociated in the flame. Like hydrofluoric acid in the previous example, EDTA behaves as a pre-flame releasing agent.

Ionization. Both the emission and absorbance of low concentrations of an alkali metal can be increased in a hot flame by adding a second alkali metal. This used to be called an enhancing interference but is in fact an example of the decrease in the degree of ionization in the presence of another easily ionized metal. When the second metal is present as an impurity, perhaps in unknown or varying amounts, the degree of suppression of ionization also varies, and the interference effect results.

As the concentration of the wanted metal increases, when the second metal is not present, the proportion of ionized atoms decreases. This, as mentioned on page 79 where calibration curvature was discussed, results in curvature of the working graph away from the concentration axis. The different degrees of ionization of potassium at two different concentrations (partial pressures) are shown in Fig. 6.

Both of these effects are eliminated by the addition of an ionization buffer. This is a second, easily ionizable, metal added in excess.

$$Li \overset{heat}{\rightleftharpoons} Li^+ + e \tag{3}$$

$$K^+ + e \rightleftharpoons K \tag{4}$$

Lithium, when present in excess, can provide sufficient electrons to prevent the ionization of potassium, although the higher alkali metals, with still lower ionization potentials, are even more efficient.

If the absorbance of an ionizable metal in the nitrous oxide flame is plotted against concentration of added ionization buffer, a plateau is usually reached–e.g., at approximately 1000 p.p.m. of potassium when added to calcium, strontium

TABLE 6.

Element	Ionization potential (eV)	% Ionization in N_2O/C_2H_2
Be	9.3	0
Mg	7.6	6
Cu	6.1	43
Sr	5.7	84
Ba	5.2	88
Al	6.0	10
Yb	6.2	20

or barium. This indicates that the ionization has been reduced to virtually zero.

The tendency to use hotter flames to increase the sensitivity of the refractory metals, and to widen the scope of atomic absorption analysis, has increased the number of ionization problems. Calcium (see Fig. 7) is more than 50% ionized at the temperature of the nitrous oxide acetylene flame, and both barium and strontium, though showing an increase in sensitivity of over twice when nitrous oxide acetylene is substituted for air acetylene, give *further* sensitivity improvement factors of about four times when potassium is present in excess as an ionization buffer.

Many metals, including aluminium[36] and silicon[310] are ionized to an appreciable extent at these same flame temperatures. Some degrees of ionization in the nitrous oxide acetylene flame are given in Table 6. The use of an ionization buffer is thus nearly always recommended with these very hot flames.

It is interesting to note that the ionization potential of lanthanum (5.61) is very comparable with that of lithium (5.39) and lanthanum therefore acts as an ionization buffer (in addition to its duties as a releasing agent) for many metals including calcium, magnesium, silicon, aluminium, etc., in the hot flames. This is partly the explanation of the versatility of lanthanum salts as spectroscopic buffers. In a complex matrix it may well be difficult to decide which particular function it is fulfilling.

Physical Causes of Interference

Incomplete Volatilization. This implies that, at the temperature of the flame, the droplets produced by the nebulizer have given rise to solid particles which, because of their high vaporization temperature, their speed through the flame, or both, are not completely converted to a vapour. The degree of atomization is therefore lower than would be expected, given the other chemical conditions in the flame. This type of interference effect is usually experienced under reducing conditions, and is then caused by the formation of metal/metal solutions of high boiling point.

Baker and Garton[43] pointed out that when volatilization of particles in the flame is incomplete, there is likely to be a noticeable departure from linearity of the absorbance/concentration relationship. A feature of this type of interference is greater curvature of calibration graphs.

The depression of the chromium and molybdenum response by high concentrations of iron in the air acetylene flame is explained by incomplete volatilization.[333] The depression has been noted previously[44,115,279] and has been reduced in the presence of ammonium chloride and aluminium chloride.

The depression is most significant in a fuel-rich flame, is less in a lean flame, and is not experienced at all in the hotter nitrous oxide acetylene flame.[390] The degree of depression increases gradually as the iron concentration is increased, levelling off only when a large excess of iron is present. This does not support the theory of the formation of a definite compound, e.g. a spinel. Furthermore, in the reverse situation, the iron absorption is little depressed by a large excess of chromium.

That this is a metal/metal effect – not an example of mixed oxide formation – is also confirmed by the severity of this interference effect in different acid matrices, which decreases in the order hydrochloric acid > sulphuric acid > phosphoric acid. This exactly parallels the ease with which iron is able to atomize from these media. If the presence of oxides were the controlling factor, the above order would be expected to be reversed.

Aspiration of chromium/iron solutions will cause relatively large solid particles, which after reduction by the flame gases consist of chromium (boiling point 2480° C) in a matrix of iron (boiling point 3000° C). These are not completely vaporized and the atomization efficiency of the chromium is low. In the reverse situation – iron in a matrix of chromium – the chromium vaporizes at a lower temperature and the iron is atomized.

A similar case is that of rhodium, the response of which in an air acetylene flame is much influenced by the presence of different acids and other metals. Addition of sodium sulphate increases the absorption above that given by a solution containing only rhodium (as rhodium sulphate or ammonium chlororhodite) and appears to overcome interferences. These interference effects also are not observed in a nitrous oxide acetylene flame.

As rhodium salts are readily reduced to rhodium metal, it must be assumed that the small solid particles obtained after evaporation of the solute contain rhodium largely as the uncombined metal, the boiling point of which is over 2500° C. There is thus incomplete vaporization and atomization. The presence of sodium sulphate prevents the formation of metallic clotlets, and as this evaporates, the rhodium is released in atomic form.

A similar explanation can be given for the releasing action of ammonium chloride, aluminium chloride and alkali sulphates on chromium in presence of iron. Instead of a solid solution of very high vaporization temperature being formed, the wanted metals are brought into solid solution in small particles of these substances. In fact, ammonium chloride and aluminium chloride would not melt, but would sublime. Sodium sulphate would also probably form sodium oxide, which also sublimes. It is likely therefore that their releasing action is a result of the rapid liberation of the metal either as an atomic or molecular vapour, both of which would make for high atomization efficiency and absorption response.

Studies on a number of cation interferences in the nitrous oxide acetylene flame, which are relevant in this context, have been made by Marks and Welcher.[250] The conclusion is reached that, though flame conditions are by no means always adequately defined in the literature, the probable explanation of many interference effects is the difference in volatility of the analyte when it is accompanied by other metal species. Most 'concomitants' actually increase the volatility of an analyte, and the majority of the interferences reported are positive. The volatility of the matrix as described by the boiling point and heat of vaporization is not always the most important consideration as the evaporation rate of small salt particles in a flame is a complex function of drop size, diffusion coefficient of the evaporating species, surface tension and heat conductivity.

Matrix Effects. These influence the number of atoms actually entering the radiation beam rather than their effectiveness once there. They usually arise from differences in the physical properties of samples and the so-called matching standards. A difference in acid concentration can, for example, cause a difference in viscosity or surface tension. The viscosity affects the rate of uptake of the sample by a given nebulizer and the surface tension affects the size distribution of the droplets formed and hence the nebulizer efficiency.

Some workers recommend that matching of standards and samples is necessary in respect of components in concentrations greater than 1% of the final solution. While this may well be sufficient for the solute, a closer match may well be necessary for the solvent, particularly where neutral sample solutions, e.g., natural waters, are to be compared with synthetic standards made up in acid medium. It is good practice to follow the same rule for both sample and standard, making both up to a stated acid concentration (0.1% hydrochloric acid, 1% nitric acid, etc.) and for accurate work to ensure that this is maintained to within 10%.

The solvent vapour pressure also directly affects the droplet size distribution both at the nebulizer, and during transport to the burner. Mixtures of solvents – particularly when low boiling point solvents are involved – must therefore be carefully controlled, as must the temperature of the solvent, particularly if samples are nebulized at temperatures near their boiling point.

Purely physical effects like those described would be expected to affect all elements in exactly the same proportion. However, this is often not the case, and some other effect is presumed also to be present. This is most probably a further result of different particle size distributions, the larger particles being incompletely vaporized. Stupar and Dawson[378] showed that some interference effects, particularly those of aluminium and silicon on magnesium, previously thought to be entirely chemical in nature, could be drastically reduced when an ultrasonic nebulizer providing a more uniform smaller droplet size was employed. The type of ultrasonic nebulizer that they employed (flow rate less than 0.1 ml/minute) may not be suitable for general analytical work, but the conclusion that good nebulizer performance characteristics and efficient spray chamber operation are essential to good analytical sensitivity and precision can still be drawn.

Scatter and Background Absorption. While these 'enhance' the actual readings obtained, the sensitivity is not improved, but a spurious absorbance is added to the true value. Scatter is analogous to turbidity in molecular spectrometry and is the result of the presence of small solid particles in the resonance beam. Such solid particles may be caused by the flame's inability to vaporize a high dissolved solids content of the sample solution, or may be due to the formation of particles, e.g. of carbon, in the flame itself.

The magnitude of this effect varies considerably with the wavelength at which measurements are being taken. For particles having a diameter of less than one tenth of the wavelength of the incident radiation, the amount of scatter according to the Rayleigh theory is proportional to λ^{-4} so that towards the lower wavelengths a very sharp increase in scatter is experienced. Light scattering therefore affects particularly those elements that absorb at the lower wavelengths, especially arsenic, selenium, zinc, cadmium and lead.

The effects of scatter are manifest in two ways. Firstly, since some of the incident light is deflected without being absorbed, a spurious increase in the absorption signal is produced. This may lead the unwary analyst into reporting results considerably higher than the true ones. Secondly, since this spurious absorption assumes the noise characteristics of the flame, an extra noise component is included in the measured signal, which, close to the detection limit of the element being measured may be greater in amplitude than the noise on the true absorption signal, or even greater than the true absorption signal itself.

The presence of the scatter effect is easily detected. All calibration curves should pass through the origin. If a particular one does not, either there is a reagent blank present in the standards or there is scatter. The amount of the scatter is evaluated in terms of its equivalent absorbance by turning away from the absorbing resonance line to another line emitted by the source in the close vicinity. This should either not be absorbed by the element being measured or should be very weakly absorbing (it does not matter whether the line originates from the cathode material or the fill gas of the hollow cathode lamp). Alternatively at the same wavelength as the absorbing line, the continuous spectrum of a deuterium lamp may be utilized. The instrument is set up again (zero energy and zero absorbance) and the sample remeasured. Any reading under these conditions is spurious absorbance and can be subtracted directly from the total reading of the resonance line in terms of absorbance. If a non-absorbing line is used it should not be more than 10 or 20 nm away from the resonance line (5 nm or less at wavelengths below 220 nm) and it should be shown also that it is non-absorbing for any other constituent of the sample. The presence of a reagent blank may be proved by comparison with purified materials, or simply evaluated by the method of standard additions.

The noise component associated with high levels of scattering is much more difficult to deal with. For this reason it is best avoided wherever possible. If it is caused by the presence of large concentrations of dissolved solids in the sample solutions, it may be necessary to separate the wanted components before atomic absorption measurement. It is good practice too, to avoid sample preparation procedures whereby high concentrations of dissolved solids are introduced. Fusions with alkaline salts can for example, often be replaced by dissolution with hydrofluoric-based acid mixtures. If the scatter originates from the flame itself, the flame conditions may have to be modified. A highly reducing flame, though an efficient atomizer, may cause too much scatter, and a compromise would have to be sought between sensitivity and noise in order to achieve the best detection limit. When the conditions permit, a more transparent flame could be used, e.g. the hydrogen argon flame is satisfactory for arsenic and selenium.

Spurious absorbance readings are also caused occasionally by absorption by molecular species formed in the flame. This particular type of interference is reduced in the nitrous oxide acetylene flame by increased dissociation of molecular species at higher temperatures. Correction is again by use of a non-absorbing line or the deuterium lamp, the latter being preferable as molecular band intensity is likely to vary much more with wavelength than a simple scattering effect.

TRACE ANALYSIS

Sensitivity and detection limit

The high sensitivity of many metals in atomic absorption suggests that this is a good technique for determining traces of metals – indeed it has been regarded by many as primarily a trace technique. In this situation, the absorbance range giving best quantitative accuracy cannot be used and the instrumentation is being operated at, or near to, its maximum sensitivity.

The analytical sensitivity of an element is defined as the concentration (quoted usually in parts per million in aqueous solution) which will absorb 1% of the incident resonance radiation of that element. Alternatively, it is the concentration giving an absorbance of 0.0044. It is, in fact the reciprocal of the slope of the calibration curve in the vicinity of the origin and its meaning is thus little different from when the term is used in other analytical techniques. Sandell [343] for example used 'sensitivity' to denote that concentration of a coloured product which in a column of solution having a cross-section of 1 cm^2 shows an extinction (*sic*) of 0.001. It will be noticed that 'one per cent absorption' is something that happens in the flame, and does not depend upon the instrument or measuring circuits employed – though a wrong reading may well be given if these are not efficient in any way.

In trace analysis one is more concerned with the limit of detection. This has been discussed at great length by many authors, but the most serviceable definition is that finally accepted by the Society for Analytical Chemistry: the detection limit in atomic absorption spectrometry is the concentration corresponding to twice the standard deviation of a series of not less than ten readings taken close to the blank level. The determination of the detection limit is then the statistical evaluation of the smallest quantity detectable with a 95% certainty. As it does not depend directly on the instrumental noise level it can be used in respect of all types of instruments including those with integrating read-out.

Detection limit is directly related to sensitivity as defined above (for if the slope of the curve is doubled the concentration corresponding to a given absorbance is halved) but is lower (i.e., better) than the sensitivity by a factor which depends on the stability and reproducibility of the read-out system.

Atomization efficiency

In order to improve the detection limit in a given analysis, it is desirable first to see if the sensitivity can be improved. This is aimed at producing the greatest possible absorbance in the flame for a given quantity of the absorbing element. When instruments employing nebulizer – spray chamber systems are being used the following points should be checked:

(i) Nebulizer take-up rate and efficiency, remembering that the efficiency of a pneumatic nebulizer is only about 10% (20% at most when one of the other devices described in Chapter 3, p. 25 is employed). If the operator is able to adjust the nebulizer himself, this is best done while the equipment is operating so that he may be sure of adjusting to give the highest absorbance. If sample volume is no problem a little greater absorbance might be obtained by using a nebulizer with an increased uptake rate, though the increase in absorbance will not be proportional and the sample or concentration noise may be increased. If the volume of sample is strictly limited, the use of one of the total sample devices referred to on page 59 et seq or an ultrasonic nebulizer, to obtain higher efficiency at lower sample uptake rates, might be considered;

(ii) spray chamber cleanliness and efficiency. The spray chamber should be clean to ensure best response time and reproducibility. It is also worthwhile to compare sensitivity and detection limit with and without flow spoilers. As suggested on p. 28, these should be optimized for a given nebulizer, and some adjustment may be desirable;

(iii) that the best flame conditions for a particular element are in use. Some guide to the best lines and flame types are given in Appendix 1.

Optical efficiency

Best analytical sensitivity is only obtained when the source produces a sufficiently narrow resonance line, and the monochromator isolates that line from all other radiation.

The quality of the resonance line from the source should be checked to ensure that it is not self-absorbed in the source lamp itself, and that it is not likely to be interfered with by other lines emitted by the source. As hollow cathode sources age, the intensity of the wanted line decreases, and that of unwanted lines may well increase. When the situation cannot be restored by decreasing monochromator slit width and increasing amplifier gain, the source lamp must be replaced.

The condition and type of the source is most important in trace analysis, for both affect the width of the emitted resonance line. Rubeska and Svoboda[340] showed that if the ratio of half-widths of emitted resonance line to absorption line is greater than one, not only is the calibration graph appreciably curved, but the sensitivity also is affected. The geometry of modern sources – hollow cathodes in particular – is arranged so that line broadening is minimized, so, for this, and the reason given in the next paragraph it is not usually possible to improve the situation by running the lamp at a lower current.

Under the conditions of trace analysis, where absorbances are low and light levels at the photomultiplier are high, the main limitation is most likely to be shot noise. The light level L given by the lamp is a function of the lamp current i:

$$L \propto i^n$$

where $1 < n < 2$. Shot noise $\propto L^{\frac{1}{2}} \propto i^{n/2}$ so that if lamp current is halved, the corresponding shot noise is reduced only by the factor 0.7–0.5. To improve the signal to noise ratio and hence the detection limit, this must be accompanied by a sensitivity improvement of at least 30–50%. As the likelihood of such an improvement in sensitivity is remote, it may well be better actually to *increase* the lamp current in order to improve the detection limit.

The level of unabsorbed light reaching the detector is reduced when the monochromator and source optics are matched to each other and optimized to the flame. Different elements may require different conditions in this respect. Calcium, for example, is atomized maximally within a quite narrow region of the flame. Sometimes, therefore, an optical field aperture is provided whereby the monochromator only receives a narrow pencil of light. The burner height is then adjusted until the maximum absorbance is obtained. The sensitivity of calcium and some other elements can thus be quite materially improved. A field aperture restricts the amount of energy falling on the detector, however, and the signal–noise ratio at the read-out stage and hence the detection limit may not be improved proportionally.

The unabsorbed light level also depends on the proximity to the resonance line of other lines emitted by the source. Their effect depends upon the resolution of the monochromator, and the slits can be closed until the limit set by resolution is reached. This is yet another adjustment which increases sensitivity at the expense of the amount of energy falling on the detector. A compromise giving the best detection limit must therefore be reached.

Noise and detection limit

The detection limit is usually a concentration of the element in solution – which we have defined on page 93 – and is thus comparable between instruments and between types of analysis. If one can improve the limit of detection of an element in simple aqueous solution, there is a good chance that it can be improved by similar measures in the presence of other substances. Another definition of limit of detection, in terms of the smallest *mass* of an element that can be detected is often used in connexion with 'total sample' techniques particularly when a solid sample can be used.

Although the detection limit is not now defined directly in terms of the noise level on the output signal, the latter affects it strongly. If the readout is in the form of a recorder trace, the position at which the analyst chooses to draw the 'average' value is less readily determined at high noise levels (see Fig. 32). Instrumental averaging or integration devices, though faithfully recording the information with which they are fed, cannot, over a short period of time, produce the true result corresponding to an infinite integration time, and hence, for a series of such readings a reproducibility which is some function of the noise level will result.

In the section on the operation of the instrument (page 74) the sources of noise were discussed, and possible means for reducing them suggested. Electronic ('Johnson') noise, lamp noise and photomultiplier dark noise are characteristic of a particular piece of equipment, and may not be readily influenced by the analyst using it. These should be low compared with photomultiplier shot noise, which as shown in the previous section, is likely to be an important source of noise in trace analysis. The effects of shot noise are increased where signals resulting from reduction of energy from the source lamp require greater amplification factors.

There is thus an optimum combination of lamp current, photomultiplier E.H.T., slit width and amplifier gain which gives the least noisy readout signal. This is ascertained in the absence of the flame.

When working near the detection limit, flame and nebulizer or concentration noise do not usually contribute a large proportion of the total noise signal, unless there is background absorbance or scatter. Thus, flame conditions should be chosen to give the maximum sensitivity, unless considerable background absorbance or scatter are thereby introduced, e.g. for elements like tin or molybdenum which, if being determined with air acetylene, require a very rich luminous flame.

Scale expansion

As the sensitivity of a given element (1 % absorption) is usually by no means the lowest concentration that can be determined, scale expansion should be used to measure such concentrations with a higher degree of confidence. The one per-cent divisions of the scale may need to be expanded to ten or more divisions.

Scale expansion in transmission instruments is usually accomplished by increasing the amplifier gain by the required factor. The unwanted part of the signal is backed off with a 'zero' control. It must be remembered that since the true zero no longer corresponds with the meter zero, absorbance values can no longer be calculated from the meter readings. This is rarely of any importance because the part of the true transmission scale actually in use is so small that it is virtually linear. It can easily be shown that the relationship between absorbance and percentage absorption is linear to 0.001 absorbance units over the range 100–90% transmission, which is the part of the scale used with a scale expansion factor of 10. As a general rule, the concentrations of elements up to 10 times their quoted analytical sensitivity will give absorption (i.e. 100 − % transmission) readings which are linearly related to concentration, and the linearity remains under conditions of scale expansion.

Scale expansion in instruments with absorbance readout is actually easier to achieve in design, and much more convenient for the operator. The result is that the actual reading on the meter or readout can be expanded by a known factor. A still linear calibration graph is obtained.

The use of scale expansion, either in trace analysis or in difference measurements (p. 111), merely permits greater accuracy of reading of the scale where small scale deflexions are involved. Both the noise and the drift associated with the output signal are also expanded by the same factor. The highest scale expansion factors are thus useful only in circumstances where an inherently stable flame and source can be employed. Elements which are atomized in air acetylene, whose resonance lines do not occur at low wavelengths and for which reliably stable hollow cathode sources are available therefore permit the highest scale expansion factors. Copper and cobalt are good examples. It should be possible, with good equipment, to scale-expand copper readings corresponding to concentrations near the sensitivity of 0.1 p.p.m. by a factor of 20 or 40, and thus obtain a detection limit of 0.002 p.p.m. or better. Cobalt, sensitivity 0.4 p.p.m., gives a detection limit of 0.01 p.p.m. in a similar way. On the other hand, such high expansion factors are of no avail for arsenic and selenium, where the detection limit is only about 4 or 5 times better than the sensitivity instead of 40 times.

Use of method of standard additions

While standard curves would be employed wherever possible, the problem of determining a trace element in an otherwise largely unknown matrix often occurs. Sufficient accuracy for trace levels is usually obtained with the method of standard additions (p. 80), as, under these circumstances, the calibration graph on either transmission or absorbance scale is likely to be linear. The method is much facilitated by the use of scale expansion, though it is necessary to check, and correct if necessary, as described on p. 92 whether or not a spurious absorbance is interfering at the resonance line being measured.

Chemical separations

The sensitivity of atomic absorption is such that many metals can be determined in trace quantities in the presence of major constituents. Nevertheless it is sometimes necessary to include a chemical separation in the preparation of the sample for atomic absorption analysis. The two main reasons for taking such a step are:

(a) when the concentration of the element to be determined is below the detection limit after normal preparation;
(b) when a separation is necessary, either from an excessive concentration of other dissolved solids (which cannot be handled by the nebulizer), or from an overwhelming chemical interference effect.

Many separation methods have been used in atomic absorption analysis. These fall into a few well-defined categories largely because of the general principle of separating, where possible, groups of elements rather than individuals, as suggested on p. 70.

Evaporation of the Solvent. This can be applied where no major constituents are present, e.g. in the analysis of water from various sources or the analysis of organic solvents. Although this method is used for concentration factors of 10–100, there is theoretically no limit to the concentration possible. The metal to be determined must not, however, be present in a volatile compound.

Deproteinization of Biochemical Samples. When trace elements are to be determined in whole blood or serum, dilution to decrease viscosity and total solids results in concentrations below the detection limit. A simple method of removing the protein is to add 1 ml of 10% trichloroacetic acid to 1 ml of the serum sample, shake and centrifuge. The supernatant liquid is then aspirated direct.

Most metal/protein bonds are broken by trichloroacetic acid, and the metals themselves, including copper, zinc and iron transfer to the supernatant.

Removal of Organic Matrix by Wet- or Dry-ashing. Dry-ashing procedures are preferred where possible, as the residue can then be taken up in the simplest possible acid solvent. Many organic matrices are completely oxidized at 550–600°C.

When volatile metals and metalloids are to be determined, wet oxidation must be used instead. Wet oxidation procedures are fully described in other publications[191] and, for some specific applications, later in this book.

Most analysts are aware of the dangers of perchloric acid in wet oxidation procedures, and it is felt by many that these are much outweighed by the advantages this acid has to offer in atomic absorption. Chief of these is that it causes interference effects with very few metals. It also allows other acid radicals to be removed easily by fuming, and thus facilitates the preparation of exactly matching simple aqueous standards.

Separation and Concentration by Co-precipitation. This often allows the exchange of a complex matrix by a much simpler one. Provided the co-precipitation system is carefully chosen, a concentration step can be achieved simultaneously. The determination of traces of strontium is facilitated by co-precipitation with added (or existing) calcium as oxalate. Zirconium and other 'group 3' metals are separated by co-precipitation on ferric hydroxide.

Removal of Matrix Metals in Metallurgical or Inorganic Analysis. This depends on the chemical properties of the matrix metal. Electrolysis may be used for copper, and also for ferrous alloys, where iron, nickel and chromium may be removed together, facilitating the determination of trace elements. Iron is also removed specifically by solvent extraction with isobutyl acetate from strong hydrochloric acid solution.

Solvent Extraction of Trace Metals. This is probably the most widely used separation technique, as it can be reduced to its simplest form. It is possible in atomic absorption, and often desirable, to extract more than one element at one operation. Specificity resides in the measuring technique. The choice of chelating or complexing reagent is therefore not limited as in colorimetry to one which gives a strong colour for the metal being determined, and complicated methods involving extractions and back-extractions in order to improve specificity are avoided.

For this reason, complexing reagents deemed unsuitable in colorimetry for their non-specificity can be used to advantage. Among these are dithizone, and the various thiocarbamate derivatives. One of the latter, ammonium pyrrolidine dithiocarbamate (APDC), first investigated by Malissa and Schöffmann (1955)[244] was reintroduced for determining traces of lead, copper and zinc by Willis[441,443] and Allan.[32, 33]

The claim that this reagent can be used to complex some 30 elements has been confirmed[419] and most of these can be extracted into organic solvents. The pH range for formation and extraction of complexes is given in Table 7.

A number of advantages result from the extraction of APDC–metal complexes

TABLE 7. pH ranges for formation and extraction of APDC–metal complexes

Periodic group	4a	5a	6a	7a	8			1b	2b	3b	4b	5b	6b
Element		V	Cr	Mn	Fe	Co	Ni	Cu	Zn	Ga		As	Se
pH for formation		2–6	2–9	2–12	0–14	1–14	1–14	0–14	1–14	2–10		0–6	2–10
pH for extraction		4–6	3–7	4–6	1–10	1–10	1–10	0–14	1–10	3–8		0–4	3–6
Element		Nb	Mo		Ru	Rh	Pd	Ag	Cd	In	Sn	Sb	Te
pH for formation		2–4	2–6		1–14	1–14	1–14	0–14	0–14	2–10	3–7	2–9	2–6
pH for extraction		2–4	1–3		1–10	1–12	1–10	0–14	0–11	2–9	4–6	2–5	3–5
Element			W	Re	Os	Ir	Pt	Au	Hg	Ti	Pb	Bi	
pH for formation			1–3	1–14	1–14	1–14	1–14	0–14	0–14	1–14	0–14	0–14	
pH for extraction			1–3	1–10	1–10	1–14	1–10	0–14	0–10	2–12	0–8	1–10	
Element	Th		U										
pH for formation	4–8		2–5										
pH for extraction	4–6		3–4										

into a suitable organic phase. The metal may be concentrated by a factor of one hundred or more. Wanted metals can be separated from high concentrations of other solutes which cause difficulties in nebulization and atomization. The atomic absorption signal for nearly all metals is enhanced by a further factor of 3–5 when aspirated in an organic solvent instead of aqueous solution.

APDC complexes are soluble in a number of ketones. Methyl isobutyl ketone which is a recommended solvent for atomic absorption allows a concentration factor of ten times, while *n*-amyl methyl ketone may be used if concentrations of up to fifty times are required. Concentrations greater than 50 times are possible using chloroform. APDC has a very high partition coefficient when extracted into chloroform, and chloroform has very low solubility in the aqueous phase. The chloroform solution, however, is not suitable for direct aspiration into an atomic absorption spectrometer and a procedure modified by wet-ashing of the chloroform extract and redissolution in 50/50 aqueous–acetone mixture is recommended.

The two procedures given by Watson[419] are as follows:

(*i*) *Extraction using methyl isobutyl ketone.*
APDC solution: dissolve 1 g of APDC in water. Dilute to 100 ml and filter before use.
To 50 ml of sample solution add 5 ml of APDC solution, and adjust to the required pH with acetic acid or caustic soda solution. The pH should be 5 except for arsenic, molybdenum, thallium and tungsten (pH 3). For manganese, raise to pH 12, mix, stand for two minutes then adjust to pH 5. With chromium and molybdenum, heat to 80°C for five minutes before proceeding. Transfer the solution to a 100 ml separating funnel, extract the complex (which may have precipitated) into 4 ml of methyl isobutyl ketone by vigorously shaking for thirty seconds, then stand for two minutes. Transfer the aqueous phase to another separating funnel and repeat the extraction with 1 ml of methyl isobutyl ketone. Discard the aqueous phase (which should now be colourless) combine the extracts in the first funnel, mix and filter through a cotton wool plug into a small beaker. This extract is aspirated in the atomic absorption spectrometer.

Standards should be prepared for all elements by the same procedure, starting with mixed aqueous standards.

If a sample volume greater than 50 ml is required to increase the concentration factor, substitute *n*-amyl ketone, which has a lower solubility in water than methyl isobutyl ketone. Watson showed that at least the following elements are quantitatively extracted with this procedure: bismuth, cadmium, cobalt, copper, iron, lead, manganese, mercury, nickel, tin and zinc, unless the limiting solubility of any complex is approached (e.g. 800 μg for mercury, and 40 μg for nickel).

(*ii*) *Extraction of smaller quantities using chloroform.*
Purified APDC solution: grind 5 g of APDC with 50 ml of acetone, collect the solid in a porosity 3 sintered glass crucible, wash with 20 ml of acetone, dry in air and dissolve 1 g of the dried APDC in 100 ml of water.
Chloroform, sulphuric acid 98% (18 M), nitric acid 70% (16 M), perchloric acid 60% (10 M) should all be of high purity.
To a volume of solution (up to 2 l) contained in a separating funnel add 2 ml of purified APDC solution and adjust the pH as the first method. Allow to stand for two minutes, then extract the complex by vigorous shaking for one minute with 10 ml of chloroform. Allow the phases to separate and transfer the chloroform extract into a 100 ml Pyrex conical beaker. To the aqueous phase add a further 2 ml of APDC

solution and repeat the extraction. Combine the extracts, add 5 ml of sulphuric acid and heat to dense white fumes. Cool, add 0.5 ml of nitric acid and again heat to dense white fumes. Repeat until a clear brown solution is obtained, then add perchloric acid dropwise until a clear or pale yellow liquid is obtained. Evaporate to about 1 ml. add 4 ml of water, cool and transfer to a 10 ml volumetric flask. Wash the beaker with two 2 ml portions of acetone, add the washings to the volumetric flask then make up to the mark with acetone. This solution is aspirated in the atomic absorption spectrometer.

Mercury, tin and bismuth are volatile under the conditions given.

Other extraction systems have been used, e.g. dithizone forms extractable complexes with about twenty metals, and the diethyl dithiocarbamates are almost as numerous as the pyrrolidene dithiocarbamates. These reagents complex many of the same metals as APDC however, and have little or no advantage.

8-Hydroxyquinoline ('oxine') is a useful chelating agent, as a number of metals form oxinates which do not react with APDC. In particular, aluminium, calcium, strontium and magnesium form extractable complexes—though not all under the same conditions—which at least allow a good concentration factor from dilute aqueous solutions.

Ion Exchange. Though invariably slower than solvent extraction, ion exchange techniques have been used to separate certain groups of metals from an undesirable matrix. Perhaps the most useful separation of this type is of trace heavy metals from higher concentrations of alkali metals as described by Biechler[57]. Especially for the analysis of industrial effluents, the method entails passage of the sample at pH 5.2 through 50–100 mesh Dowex A.1. resin in a 10×1 cm column. Copper, lead, zinc, cadmium and nickel are retained with 1 litre samples, but complete retention of iron requires a sample of less than 500 ml. These metals are eluted with 25 ml of 8.0 M nitric acid and the eluate made up to 50 ml. The reference solution should be 4.0 M nitric acid.

Phosphate may be separated from solutions containing calcium and strontium on a de-acidite FF (acetate form) column like the one described by David[113] though the need for this particular separation is probably lessened in the light of hotter flames and knowledge of releasing processes.

Two chelating resins, Chelex 100 and Permutit S1005, were investigated by Riley and Taylor[323] and were found to retain a number of elements, including the rare earths. Bismuth, cadmium, cerium, cobalt, copper, indium, lead, manganese, nickel, scandium, thorium, yttrium and zinc were completely recovered by elution with mineral acid (2M) and molybdenum, tungsten and vanadium by elution with ammonia (4M). Rhenium and silver were 90% recovered.

Addition of miscible non-aqueous solvents

The property of acetone, when added to an aqueous solution, of increasing the absorbance of some metals has been known for some time. Advantage has been taken of this in the recommended method for determining iron in serum (p. 168). An analytically useful increase in absorbance is only obtained if the true 'enhancement' factor is greater than the dilution factor. The behaviour of methanol, ethanol and acetone in this respect has been investigated by Panday and Ganguly[290]. Such useful increases in sensitivity were found for 1:1 dilutions with acetone for copper, manganese, iron, rubidium, calcium and magnesium. Neither methanol nor ethanol were sufficiently effective.

With lower proportions of acetone, the magnitude of the effect is too small. With greater proportions the dilution is too high and instability of the flame caused by vaporization in the nebulizer may also be experienced.

DETERMINATION OF MAJOR COMPONENTS

The high sensitivity of many elements has led many workers to maintain that atomic absorption spectrometry is essentially a trace technique and is not to be considered for the determination of higher or major concentrations.

Provided that stable and reproducible instrumental performance can be established and that as much analytical care is taken as would be expected in volumetric or gravimetric assays, there is no reason why coefficients of variation of 0.5% and better should not be achieved. It is not suggested that atomic absorption should henceforth replace many of the reliable classical assay methods. It is believed, however, that many analyses which are difficult and time-consuming by classical methods, and not sufficiently reliable by newer physical methods can be tackled by atomic absorption with advantages in speed and convenience, and with sufficient accuracy for all except perhaps the most stringent assay requirements. Examples of such analyses occur in complex oxide materials—slags, cements, glasses, rocks, ores etc.—some of the more complex alloys, plating solutions and also wherever metal-containing additives have to be controlled in an essentially organic matrix.

The achievement of high accuracy in atomic absorption determinations is both an analyst's and an instrumental problem.

Sample preparation and calibration

The accuracy of an analytical result cannot be better than the accuracy of the methods used to prepare the sample itself. Random errors are statistically additive, and they must therefore be minimized throughout the whole procedure. The weighing of the original sample and the dissolution process must be carried out with the same care and refined technique employed in volumetric or gravimetric assay. Subsequent dilutions must be performed in high quality volumetric ware.

The accuracy of calibration standards is a *sine qua non* in this type of analysis, but the question of interferences must also be carefully examined so that standards and samples can be appropriately matched. Even minor interferences are now important. This problem may be circumvented if chemically analysed standard samples of identical type are available. These may be obtained in the U.K. from the Bureau of Analysed Samples and in the U.S. from the National Bureau of Standards. Many types of alloy, rocks, ores etc. are included in their catalogues. These standards may also be used, as in the author's laboratory, to check that all interferences have been allowed for in the making-up of independent synthetic standards, thus improving confidence in the final results.

Releasing agents and ionization buffers must, of course, be added in equal quantities to both samples and standards. The effect of all the other major constituents of the sample on the one being determined may have to be examined. The materials from which standards are prepared may also have to be chosen carefully, for example, if an alkali salt such as potassium dichromate is used to prepare high concentration standards, the ionization buffering effect of the potassium, present only in the standards, may cause low results to be returned for the samples. Such effects can usually be foreseen, and if not preventable, can often be overcome by adding a small excess of the offending constituent to both samples and standards.

Because most of the best-known resonance lines of the elements are extremely sensitive either the preparation of the analysis solution must incorporate considerable further dilution steps, or other means must be sought to decrease the sensitivity so that the measured absorbance falls within the instrumental range of greatest accuracy. Large dilution factors are not ideal as they incur manipulative errors and an unnecessary increase in the time spent on the analysis.

Decreasing analytical sensitivity

The alternative is to decrease the absorption signal in the instrument itself. This must be achieved without loss of stability and reproducibility which therefore precludes modifications to the nebulizer or its associated gas flows. Flame conditions and observation height, too, should only be adjusted to give the best stability and freedom from noise.

Sensitivity can be decreased by choosing another absorbing line of lower absorptivity. Most elements have sufficient resonance lines within the effective instrumental wavelength span to allow a wide range of concentrations to be determined. The use of more than one line per element, of course, is normal practice in emission spectroscopy. Useful alternative lines are quoted, together with an

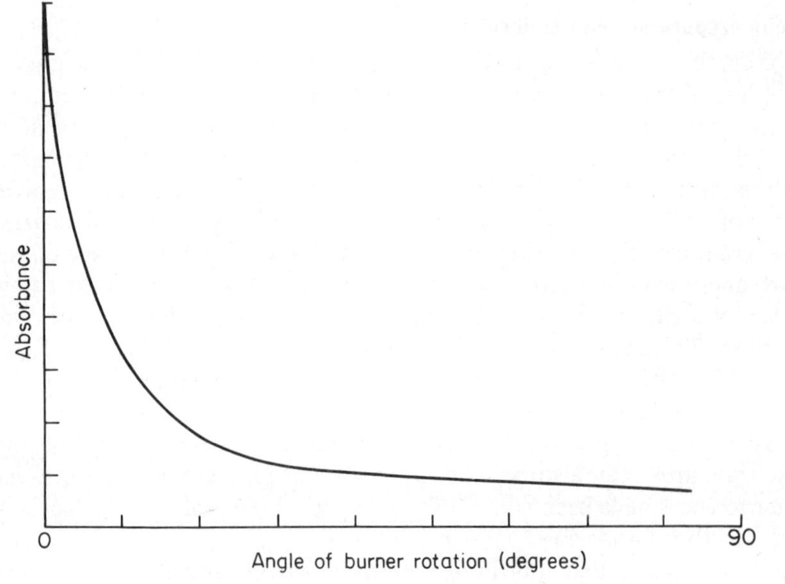

FIG. 36. Effect of burner rotation on sensitivity.

estimate of their sensitivities in the information on individual elements in Appendix 1. It should be noted, however, that the most sensitive line is usually also the strongest line and therefore higher amplifier gain factors are nearly always required for the alternative lines. A practical point, too, is to ensure that the alternative line is in a part of the spectrum transmitted by the particular source lamp in use. The alternative line for magnesium at 202.5 nm for example is not transmitted by a lamp with a 'UV glass' window.

Sensitivity is also effectively decreased by rotating the long path length burners through an angle about their vertical axis, so that the resonance beam passes through only a narrow section of the flame. The rate at which the absorbance decreases with the angle of rotation (see Fig. 36) is obviously very high at small angles of rotation and thus it is difficult to set the burner angle accurately to ensure a decrease of sensitivity by a given factor. It is probably best to use the burner only fully angled (i.e. at 90°) from its intended position, when about one eighth of the normal sensitivity is obtained. Rotation of the burner is clearly preferable to the use of an alternative resonance line if the latter requires gain factors to be increased to the point where noise on the output signal becomes noticeably greater.

Where a circular Meker head is provided for emission measurements, this will give an effective path length of about one fifth that of the 10 cm air acetylene burner, with a sensitivity decrease in approximately the same ratio. Note, however, that the air acetylene emission burner cannot be used safely with nitrous oxide acetylene, nor will an air propane emission burner support an air butane flame.

Instrumental performance and reading accuracy

If an analytical accuracy of 0.2 % is to be achieved, the stability and reproducibility of both zero absorbance base-line and measured absorbance signal must also be of this order.

Observable drifts in the base-line or resonance line absorptivity cannot therefore be allowed during the period of the actual analysis, i.e., when samples and standards are being aspirated. While a double-beam spectrometer corrects for variations in source lamp intensity, it may not reveal absorptivity drifts due to lamp warm-up or alteration in flame conditions, and so, whatever instrument is in use, it is wise to check for drifts originating in any part of the system by repeating one of the high calibration standards at frequent intervals, perhaps, after every four or five samples, and certainly at the end of the analysis.

Good reproducibility requires, in addition to general stability, an output signal with the least possible associated noise. The origins and suppression of noise were discussed (p. 74 et seq.) and as lack of sensitivity is rarely a problem in this type of analysis, the instrument conditions can be chosen with low noise as the primary objective.

The part of the instrument read-out scale in which the greatest reading accuracy is obtained depends on numerous factors, but as pointed out by Weir and Kofluk,[422] in the most precise work, absorbance readings must be limited to a relatively narrow range. They demonstrated that coefficients of variation usually have minimum values for absorbance between 0.5 and 0.8, (see Fig. 37) when the instrument is set up to give best precision.

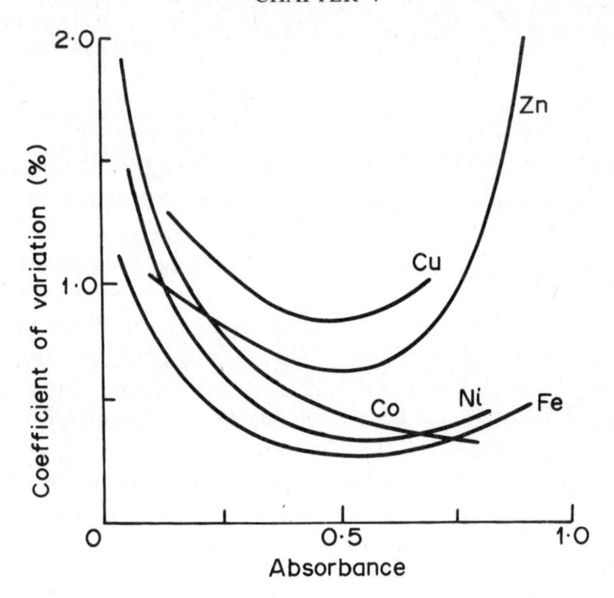

FIG. 37. Coefficients of variation at different absorbance levels reported by Weir and Kofluk.[422]

The optimum absorbance at which maximum photometric accuracy can be obtained may be derived from consideration of the signal/noise and other instrumental characteristics.

In the following, ΔA and ΔT are random errors associated with a single reading in the absorbance and transmission readout modes respectively.

By definition
$$A = -\log_{10} T = -\log_e T \, . \, \log_{10} e \qquad (1)$$
so
$$dA = -(\log_{10} e) \frac{dT}{T}$$
or, for finite increments
$$\Delta A = -(\log_{10} e) \frac{\Delta T}{T} \qquad (2)$$

Below are listed the sources of random error, together with their associated error function (i.e., the way in which the incurred error is a function of the readout signal):

1st Category $\Delta T \propto T$
- (i) Source lamp fluctuation
- (ii) Gain drift
- (iii) Optical stability
- (iv) Flame scattering effects
- (v) Readability of linear absorbance readout scale

2nd Category ΔT is constant
- (vi) Electronics (Johnson) noise
- (vii) Readability of linear transmission readout scale
- (viii) Flame emission noise breakthrough

3rd Category $\Delta T \propto T^{\frac{1}{2}}$ (ix) Photomultiplier shot-noise

4th Category $\Delta A \propto A$ (x) Variations in analyte concentration (i.e. nebulizer noise)

(xi) Variations in absorptivity coefficient (e.g. atomizer noise)

(xii) Variations in cell path length (i.e. flame 'flicker').

The first three categories are common to all types of absorption spectrometry and were summarized by Ford[143]. The fourth type, characterized by Roos[329] as $\Delta T \propto T \log_e T$, arises essentially from the dynamic nature of the flame absorption cell.

The reason for the particular error function ascribed to each error source will in most cases be obvious. In (i)–(iv), variations directly affect the amount of energy falling on the detector. In (v), ΔA is constant, so, from equation (2) $\Delta T/T$ is constant, so $\Delta T \propto T$. In (vi), (vii) and (viii), ΔT is clearly constant for a given instrument setting. Johnson noise will be greater, however, at high gain settings (low source intensity) but is much less likely to be a precision limiting factor than shot noise. Shot noise itself (ix) results from the quantized nature of electromagnetic radiation and the relationship $\Delta T \propto T^{\frac{1}{2}}$ is proved by statistical theory.

If analytical precision S be defined as $C/\Delta C$, then, when Beer's law is true.

$$S = \frac{C}{\Delta C} = \frac{A}{\Delta A} \tag{3}$$

Combining this with equations (1) and (2)

$$S = \frac{A}{\Delta A} = \frac{T \log_e T}{\Delta T} \tag{4}$$

For category 1 error functions, where $\Delta T \propto T$, from equation (2) ΔA is constant thus, $S \propto A$. Hence for all errors in this category the precision should increase directly as the absorbance increases. There are of course practical limits to this which are dictated by the presence of stray light and other deviations from Beer's law.

In category 2, where ΔT is constant, from equation (4)

$$S = \frac{T \log_e T}{k}$$

So
$$k\frac{\mathrm{d}S}{\mathrm{d}T} = 1 + \log_e T = 0 \quad \text{for maximum } S$$

therefore
$$\log_e T = -1$$

and from (1)
$$A = \log_{10} e = 0.43$$

This is the well known derivation of this value for absorption spectrometers where the factor limiting accuracy is the readability of the T scale. In atomic absorption this error function is more likely to apply only in the case of serious flame emission noise breakthrough, when precision would, in any case, be relatively low.

In category 3, substituting $\Delta T = k'T^{\frac{1}{2}}$ in equation (4)

$$S = \frac{T \log_e T}{k'T^{1/2}} = \frac{T^{1/2} \log_e T}{k'}$$

Now, $$k'\frac{dS}{dT} = T^{-1/2}(1 + \tfrac{1}{2}\log_e T) = 0 \text{ for maximum } S$$

Hence $\log T = -2$, and from equation (1) $A = 0.86$.
This optimum absorbance value is usually quoted for spectrometers with photomultiplier detectors, where reading accuracy is limited by shot noise. In cases where highly stable instrumental and flame conditions can be achieved, this would be the limiting factor in atomic absorption spectrometry too.

In category 4, which is probably most serious in atomic absorption, $\Delta A \propto A$. so $S = A/\Delta A =$ constant [and it follows from equation (4) that $\Delta T \propto T \log_e T$] so the precision is independent of the absorbance, the transmission and concentration.

These results are simply interpreted for instruments with transmission readout. Category 1 fluctuations are proportional to the scale reading and thus, within the limits of calibration linearity, precision becomes greater toward low transmission values. In category 2, best precision is around T values of 37% and in category 3 around 14%. In category 4 precision is independent of T.

Deviations from Beer's Law. The above is purely an idealized approach to the error problem, for, not only do errors from all four categories operate simultaneously, but deviations from Beer's Law considerably modify these results. The effect of such deviations can be calculated assuming that they are largely the result of unabsorbed (and unabsorbable) light reaching the detector.

If s is unabsorbable and l is absorbable then

$$(s + l)\,T = s + l.10^{-abC}$$

where T is observed transmittance and abC is true absorbance.

i.e., $$sT + lT = s + l.10^{-abC}$$

so $$10^{-abC} = \frac{s - sT - lT}{l}$$

hence, $$abC = \log_{10}\frac{(l + s) - s}{T(l + s) - s}$$

$$= \log_{10}\frac{(1 - B)}{(T - B)} = \log_{10}\mathrm{e} \,.\, \log_e\frac{(1 - B)}{(T - B)}$$

where $$B = \frac{s}{l + s} \text{ i.e. the proportion of stray light.}$$

Differentiating $$ab\,.\,dC = -\log_{10}\mathrm{e}\,\frac{dT}{(T - B)}$$

so $$S = \frac{C}{\Delta C} = \frac{C}{dC} = -\frac{(T - B)}{\Delta T}\,.\log_e\frac{(1 - B)}{(T - B)} \quad (5)$$

The four error functions can now be substituted in this equation. By judicious choice of the value of the constant k for each case, the relationship between precision and absorbance values for various values of B (i.e. for different degrees of variation from Beer's Law) can be calculated.

For category 1 errors, $\Delta T = kT$ so equation (5) becomes:

$$S = \frac{(T - B)}{kT} \cdot \log_e \frac{(1 - B)}{(T - B)} \tag{6}$$

If, by way of example, the source of the worst error is the absorbance scale, which should be readable with an accuracy of $0.001A$, $\Delta T \approx 0.002T$. Fig. 38(a) shows the values of S plotted against A for the values of B: 0, 5, 10, 20, 30, 40 and 50 %. As already derived, when $B = 0$ (i.e. Beer's Law is obeyed) the relationship is linear, and precision increases with the absorbance reading. However, when stray light is present, the precision is seen to pass through a maximum. At 5 % stray light the maximum precision $S_{(max)}$ is at absorbance 0.9, and for higher values of B, the maximum precision occurs at lower values of A.

Note that the slopes of the curves at any point are directly proportional to $1/k$, so if k is greater than 0.002 the values on the ordinate may be reduced in proportion. For a serious flame scatter interference, k could be 0.01 or greater. Thus, while for a given value of B, the absorbance at which maximum precision occurs may be expected to remain the same, the precision itself is less (i.e. the error is greater) in proportion.

It would be hoped that in a good modern atomic absorption spectrometer, the other sources of category 1 error (viz., source lamp fluctuation, gain drift and optical instability) would be small compared with the examples given.

Category 2 errors where $\Delta T = k$ give the basic equation

$$S = -\frac{(T - B)}{k} \cdot \log_e \frac{(1 - B)}{(T - B)} \tag{7}$$

Considering the readability of the T scale, ΔT should not be greater than 0.1 % of T, $k = 0.001$.

Relationships are plotted in Fig. 38(b) assuming this value. As already calculated, S_{max} for $B = 0$ occurs at 0.43, but, as B increases above zero, $S_{(max)}$ occurs at progressively lower absorbance values. The maximum precision values are themselves relatively low, and this suggests a real disadvantage of instruments which read out linearly in transmission, particularly when higher concentrations are being determined.

In linear absorbance instruments, the worst source of error in this category would be emission noise breakthrough. This has been experienced by several workers when attempting to determine, for example, barium in presence of high concentrations of calcium. The value of k may then rise to between 0.01 and 0.1. As the slopes of the curves are proportional to $1/k$, with such values, very low precisions are indicated, particularly at low absorbance values and when there is deviation from Beer's Law.

Category 3 errors, due to shot noise, where $\Delta T = kT^{\frac{1}{2}}$ produce the basic equation

$$S = -\frac{(T - B)}{kT^{1/2}} \cdot \log_e \frac{(1 - B)}{(T - B)} \tag{8}$$

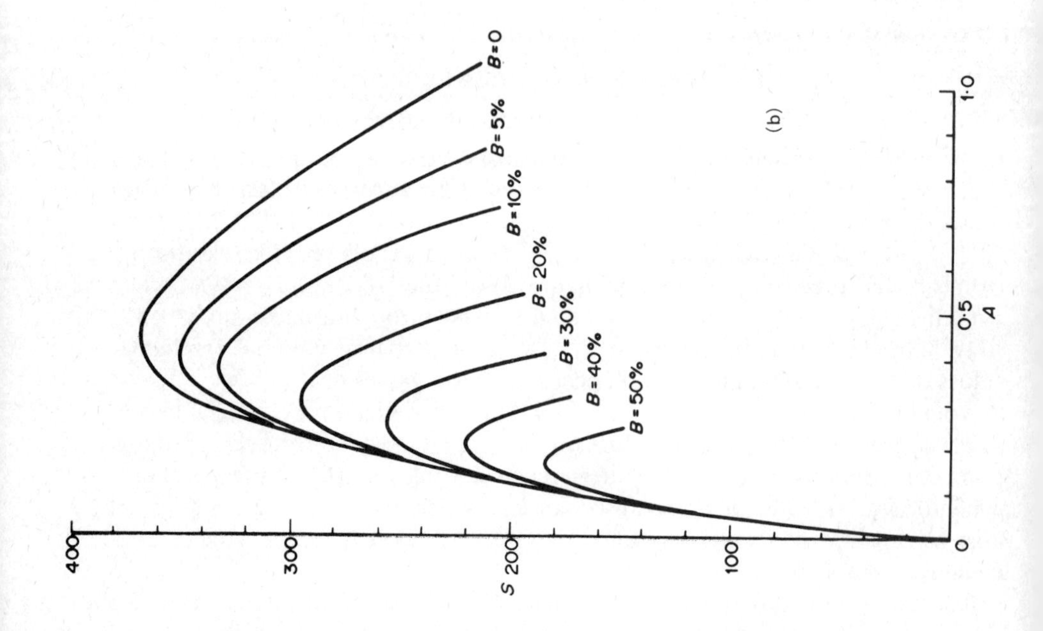
(b)
S
A
0
100
200
300
400
0·5
1·0
B = 0
B = 5%
B = 10%
B = 20%
B = 30%
B = 40%
B = 50%

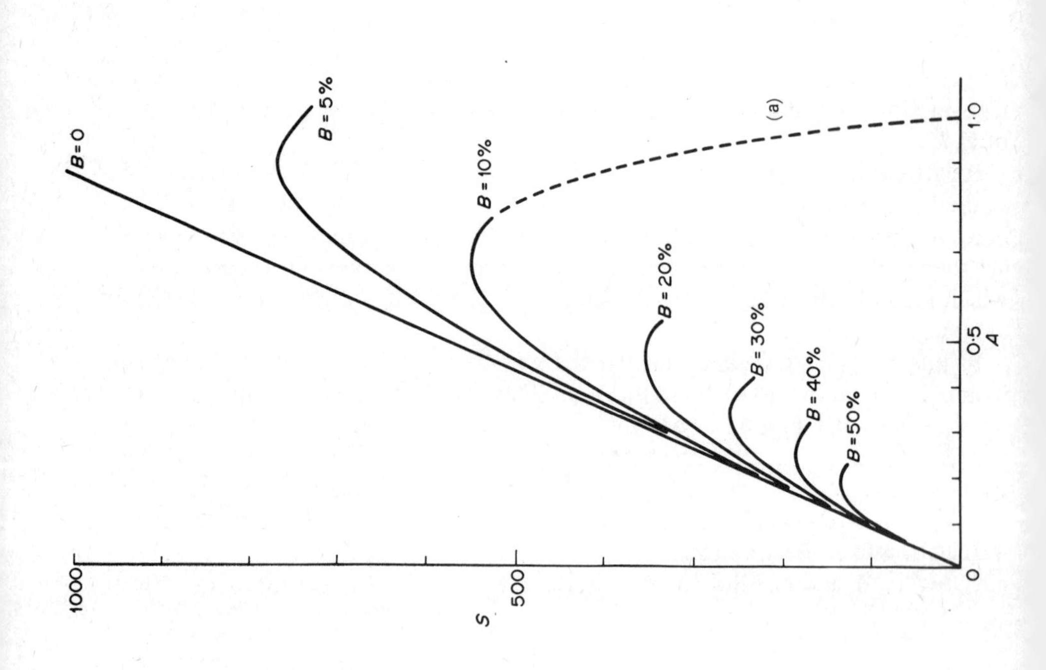
(a)
S
A
0
500
1000
0·5
1·0
B = 0
B = 5%
B = 10%
B = 20%
B = 30%
B = 40%
B = 50%

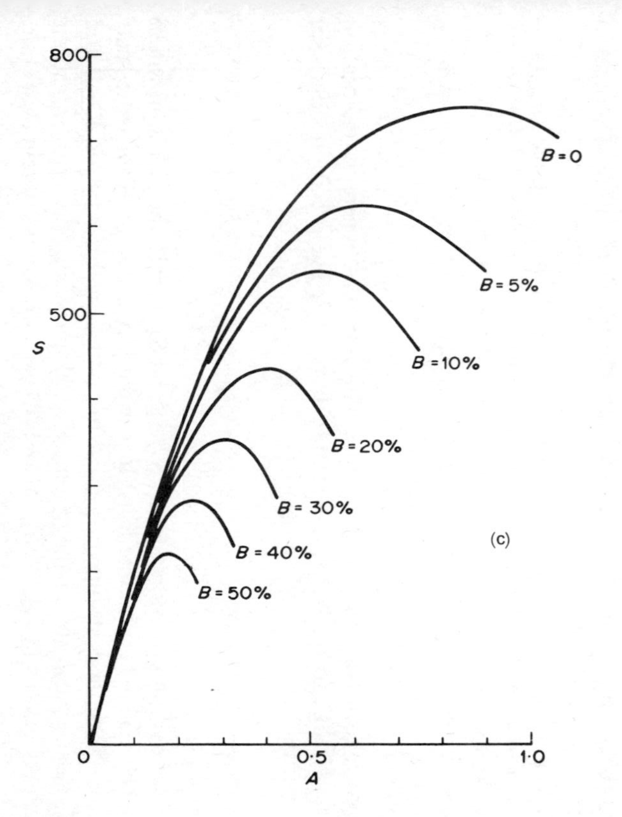

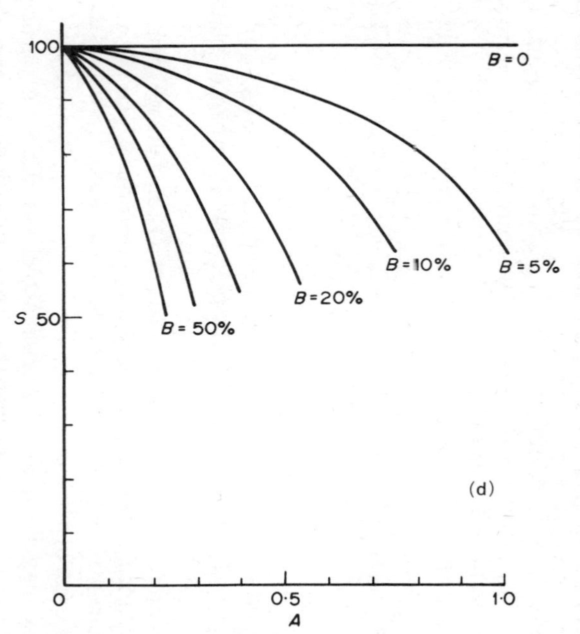

FIG. 38 (a) Plot of precision against absorbance for category 1 error function, $\Delta T = 0.002\ T$. (b) Plot of precision against absorbance for category 2 error function, $\Delta T = k = 0.001$. (c) Plot of precision against absorbance for category 3 error function $\Delta T = 0.001\ T^{\frac{1}{2}}$. (d) Plot of precision against absorbance for category 4 error function. $\Delta A = 0.01\ A$ i.e., $\Delta T = 0.01\ T \log_e T$.

With a modern photomultiplier in use under average conditions in an atomic absorption spectrometer, shot noise should be such that k is not greater than 0.001. Fig. 38(c), in which this value has been used, confirms that, when $B = 0$, $S_{(max)}$ occurs when absorbance is 0.87. Again, as deviations from Beer's Law increase, $S_{(max)}$ occurs at lower absorbances, e.g., for 5% of stray light the maximum precision is at about 0.65 absorbance units.

A somewhat different form of relationship is shown by the category 4 error function.

The basic equation becomes

$$S = -\frac{(T - B)}{kT \log_e T} \cdot \log_e \frac{(1 - B)}{(T - B)} \tag{9}$$

If total atomization noise produces a possible reading error of 1%, i.e., $\Delta A = 0.01A$, then from equation 4, $\Delta T = 0.01\ T \log_e T$. The precision/absorbance relationships, using the value 0.01 for k, are plotted in Fig. 38(d).

As deviations from Beer's Law are introduced, precision falls off markedly with increased absorbance. The comparatively low precision values are also a feature of this error category, which is probably the greatest source of inaccuracy in the conventional atomic absorption technique.

These relationships may be used to deduce actual precisions in particular practical cases but it must be realized that they apply to a single instantaneous reading. Where the effect of the source of error is to introduce a noise component on the output signal, this can be reduced by taking averaged readings, or by integrating, both over a selected period of time.

Nevertheless, computed precision values are useful in two particular ways:
(i) they enable an operator to compare directly the effects of different sources of error upon the final result; for example, it is often assumed that photomultiplier shot noise is the main limitation on precision when the flame is stable and transparent. The foregoing results show that, in the case of an instrument with transmission readout, this may well not be true;
(ii) they enable the effects of different degrees of variation from Beer's Law to be calculated.

The value of 10% for B is not unrealistic or unusual when a resonance line occurs close to or in a group of non-absorbing lines. This is the case where the calibration graph is asymptotic to the horizontal at absorbance 1.0. The precision values are the inverse of the errors, which are statistically additive. So, for example:

$$\frac{1}{S^2} = \frac{1}{S_1{}^2} + \frac{1}{S_2{}^2} + \frac{1}{S_3{}^2} + \frac{1}{S_4{}^2}$$

where S is the overall precision and S_1, S_2, S_3 and S_4 refer to the precision values relative to the various error functions. If just the errors due to absorbance scale reading, shot noise and atomization noise are added in this way for $B = 0$ and $B = 10\%$, at a typical absorbance value of 0.50 and using the error constants from which the graphs in Fig. 38(a), (c) and (d) were drawn, we obtain the result:

for $B = 0$ $\quad S = 97.4$

for $B = 10\%$ $\quad S = 82.8$

and these are the expected precisions of a single reading in these two cases (these would be popularly expressed as errors of '1 in 97.4' and '1 in 82.8' respectively).

Conclusions from Treatment of Error Functions. The value of the constant k chosen for use in the calculations which gave the graphs in Fig. 38 are typical of those encountered in traditional atomic absorption techniques. The different error functions contribute different magnitudes of error (roughly in the descending order: categories 4, 2, 1 and 3) though these are directly affected by the chosen values for the constant k.

The importance of stable flame conditions is thus again emphasized, as improvements here will generally contribute the greatest gain in repeatability of results and therefore in overall precision.

All error sources operate simultaneously to a greater or less extent. It would therefore appear that, for low deviations from Beer's Law, taking the relevant graphs into consideration, the maximum precision is most likely to occur between the absorbance values 0.5 and 0.8. This observation agrees with the practical data given by Weir and Kofluk[422] which has already been mentioned.

Difference measurements in atomic absorption

Difference measurements, by which is understood the use of scale expansion and zero back-off of the absorbance scale, so that a calibration does not involve a 'zero' standard, can be used to improve the precision of reading at comparatively high absorbance readings. Thus it is possible to improve precision and accuracy of determination at the higher concentration levels.

The method was discussed by Kahn[194] for avoiding high dilutions and by Welz[425] in relation to the determination of major constituents in metals.

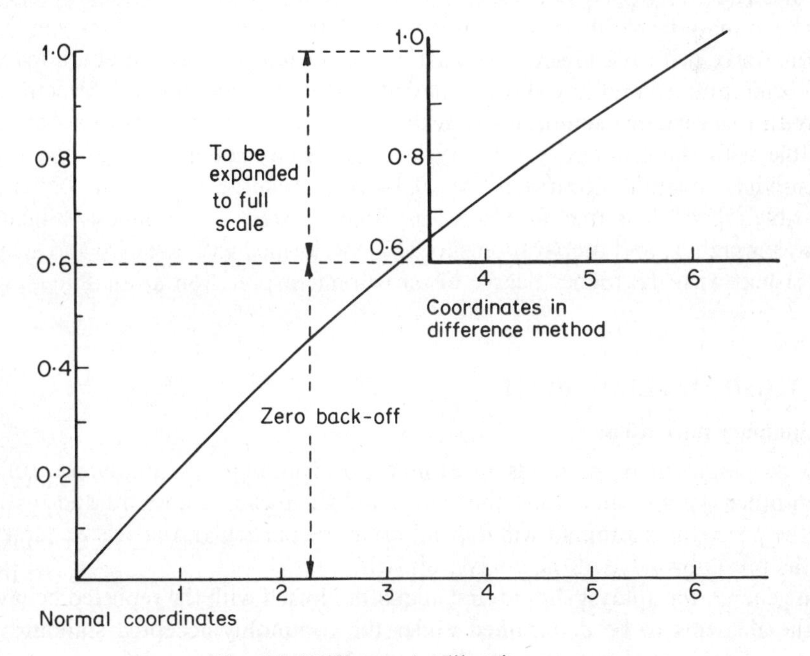

FIG. 39. 'Difference' calibration.

In principle, the calibration range is limited to bracket the range expected in the samples, and dilution and sensitivity are adjusted so that readings fall somewhere within the normally accepted range of optimum accuracy. The function of the zero standard is now assumed by the standard at the low end of the calibration range. The situation is shown diagrammatically in Fig. 39.

From the foregoing sections it is clear that for best precision of results the true absorbance values should fall within the range of 0.5–0.8.

In order to obtain maximum benefit from difference readings, the spectrometer readout controls should allow zero back-off to at least 0.7 absorbance units, and provide a scale expansion of ten times. This would allow the range 0.7–0.8 A to be read across the full scale of the readout meter or other device. Thus the precision of readings of the absorbance scale might be improved by a factor of 10, or by whatever scale expansion factor is employed. This can contribute a real improvement to the accuracy of analysis where a meter reading or recorder trace is involved because it will minimize the effects of parallax on the former and of the pen thickness on the latter.

Operating Procedure. The instrument should first be set up in the conventional mode incorporating the optimum or recommended settings. If the concentration of the highest standard for a particular element is more than 200 × sensitivity (i.e. it gives an absorbance greater than 1.0) the sensitivity must be reduced by one of the means already described (page 102).

The lowest standard of the adopted range is then aspirated, and the resulting absorption signal is brought to zero absorbance on the readout scale by increasing the amplifier gain. Next, the highest standard is again aspirated, and sufficient scale expansion is introduced to give a reading on the readout scale between 0.8 and 1.0.

The remaining calibration standards and the samples are then run in the normal way, with the lowest standard between each, so that the operating baseline is maintained and any short-term drift effects are immediately detected.

With this type of technique, given that good stable flame conditions are compatible with the required sensitivity level (which is not necessarily the highest sensitivity) overall accuracies of 0.5–0.2% relative, or even better, are possible.[389, 390] It is therefore necessary that all steps in the analysis, including sample weighing and preparation should be performed with the care and accuracy consistent with the higher degree of measurement precision attainable.

METHOD DEVELOPMENT

Preliminary indications

The development of methods in atomic absorption is essentially no different from other types of analytical chemistry, and there can be no hard and fast rule, for the procedures adopted will depend upon the particular analysis in hand, and on the pre-knowledge of the analyst himself.

In general, the analyst should first acquaint himself with the reported behaviour of the elements to be determined under the commonly accepted standard conditions. These are summarized in Appendix 1.

If the sample is a dilute aqueous or non-aqueous solution of the elements in question, then a comparison is made between the reported sensitivity, and concentration expected to be present in the sample. If the sample is solid material, it must first be assumed that it will be diluted by a factor of 100 × on being taken into solution.

For best quantitative accuracy the concentration in the prepared solution should, as shown in the previous section, be between 20 and 200 times the sensitivity. To achieve this, total dilution factors of up to $1:10^4$ are regarded as permissible. If higher dilutions should be required, the sensitivity also should be reduced (see p. 102).

If the concentration lies between 1 × and 20 × the sensitivity, the situation can either be accepted as a 'trace' analysis, when a result of the correct order but with a lower precision can be reported, or steps may be taken to increase the concentration in order to maintain precision.

For concentrations below the sensitivity, the procedure depends upon the noise level and attainable detection limit, but a concentration step is made unless a purely semiquantitative indication is all that is required.

Calibration

With a series of standard solutions covering each concentration range expected, the flame and instrumental conditions should be adjusted to give best sensitivity or best stability as required. The effects of the addition of other major constituents are then checked. Interference effects are then countered by use of the appropriate type of spectroscopic buffer and/or by modification of the flame conditions. It may be necessary, for example, to employ a hotter flame and an ionization buffer, or satisfactory accuracy may be achieved simply by careful matching of the standards to the samples. The latter expedient is usually satisfactory only as long as the concentrations of the major constituents remain virtually constant, and provided they do not exert too great a degree of depression.

The calibration and general procedure is then checked against standardized samples, if available, of the same type as those to be analysed. If chemically analysed standard samples are not obtainable, it is necessary to establish the degree of recovery of accurately known amounts of the wanted elements added to the sample solutions. Provided the method development and analytical procedure have been carefully carried out, recoveries should be within $\pm 2\%$ (relative) of the amounts added. This would normally be sufficient to establish the validity of a method where the sample and standard solutions actually being measured contain mixtures of inorganic ions. In the presence of organic material such as protein, the metal already existing in the sample may be bound differently from that added subsequently and give rise to a different response in the flame. In such cases the results obtained with the organic matter present must always be compared with those given after the organic material has been destroyed by dry- or wet-ashing.

Precision

An indication of the accuracy expected from the developed method is always of value and is essential when different methods are to be compared. Errors may be random or systematic. Systematic errors almost invariably indicate a difference in chemical constitution between prepared sample and standard solutions which should have been taken into account during the method development investigations. In principle, therefore, it is always possible for systematic errors to be eliminated. Random errors cause poor repeatability between results obtained in one laboratory and poor reproducibility between those obtained in different laboratories.

A measure of the precision is therefore obtained with a reproducibility test in which the standard deviation or coefficient of variation is calculated from at least ten and preferably twenty replicate results. To check the instrumental reproducibility, the test is performed on one prepared solution, a blank or pure solvent being aspirated between successive readings. Finally the test should be carried out on the same number of individually prepared solutions from the same original bulk sample, in order to show up errors originating from the sample itself (heterogeneity) or the preparation procedure.

In a good quantitative procedure. coefficients of variation of $\pm 1\%$ or even better are obtainable, so that the precision can be improved by replication. If n replicate samples are analysed, the precision is $1/n^{\frac{1}{2}}$ times the coefficient of variation. This will also be the accuracy of the method in the absence of systematic errors.

Chapter 5

The Applications of Atomic Absorption Analysis

In view of the considerable literature that now exists on atomic absorption analysis, it would require a very large volume indeed to discuss in any detail all the known applications. Wherever metals are determined, and in whatever matrix, atomic absorption can be applied. There is very little difference in the methods required to handle the materials within certain well defined groups. One such group includes waters from various sources, wines and beers. Another includes food products, feedingstuffs, plant materials and biological tissues, and a third contains various complex oxide mixtures such as slags, rocks, ores, cements, glass and ceramics. When prepared for analysis, the solutions contain much the same elements and therefore the interference problems are also similar. To differentiate between 'industrial' and 'geochemical' applications, or even between 'industrial' and 'biochemical' is often pointless.

When seeking information on a particular analysis, therefore, a worker may need to look beyond the confines of his own field, as identical problems may have been solved elsewhere.

We attempt to divide the whole analytical scene broadly into metallurgical, inorganic, organic, and biochemical sectors, though even here some overlapping must occur.

Unless special details are given, the reader should refer for element sensitivities and generalized operating conditions to the tables in Appendix 1.

THE ANALYSIS OF WATERS AND DILUTE AQUEOUS SOLUTIONS

This topic is treated individually as all waters, and some types of solid sample after preparation, are, in effect, dilute aqueous solutions. Although waters might be considered to be the simplest possible samples for atomic absorption, there has nevertheless been a large amount of work reported, and this falls in three main areas – purified water, river water and industrial effluents, and sea water. The elemental sensitivities and detection limits published in the scientific and the manufacturers' literature usually relate to aqueous solutions. These are therefore the levels of performance expected in water analysis. Considerable use has also been made of the three main concentration techniques – evaporation, solvent extraction and ion-exchange.

Purified water

Techniques for determining zinc, lead, copper, nickel, calcium, magnesium, iron and manganese were described by Boettner and Grunder.[64] The APDC-MIBK method was used by Fishman and Midgett [141] for cobalt, nickel and lead, the extraction being carried out at pH 2.8. In an examination of some Indian water supplies, Soman et al.[367] used dilution and concentration by evaporation as appropriate to determine cobalt, chromium, copper, iron, lithium, manganese, rubidium, strontium and zinc. The more stringent requirements of the steam water circuits of power stations were described by Wilson[449] who used the APDC-MIBK extraction as necessary, though direct analysis gave detection limits in μg/l of: iron 30, copper 15, zinc 15, nickel 60, calcium 10, magnesium 1 and sodium 4.

Most purified waters can therefore be analysed without prior chemical treatment, but where levels below their detection limits are to be determined, one of the concentration procedures – particularly the APDC-MIBK extraction – must be employed, as described on page 98.

Rivers and industrial effluents

Unhappily it is a sign of the times that these should be grouped together in the eyes of many analysts. The essential differences from purified water are the presence of suspended organic matter, and the increase in concentration of many – usually toxic – elements.

Many river authorities and water boards employ atomic absorption in water pollution control. The methods are usually similar to the procedure suggested by Price.[305]

Suspended matter in the sample is first separated, or, if it contains any of the elements to be determined, it must be homogenized and dissolved or extracted with nitric acid. In extreme cases, the solid matter is centrifuged off and treated separately as a sludge. Sludges containing organic matter are wet- or dry-ashed like other organic matrices (see page 97) and taken up in mineral acid. If they contain siliceous material the silica is removed by treatment with hydrofluoric and perchloric acids. Clear or clarified water is acidified to about 1% with nitric acid and can be aspirated in the atomic absorption spectrometer without further treatment if the concentration ranges present can be handled. Calibration standards made up as described earlier should also be acidified 1% with nitric acid. If concentrations are too high, they are brought within the best range by dilution with deionized water also acidified with 1% of nitric acid. Calcium and magnesium, to relate to hardness values, are determined in this way. If concentrations are too low, the sample may be evaporated to a lower volume, or the wanted metals extracted into an organic solvent as described on page 98. The ion-exchange technique also referred to there was used for the analysis of industrial effluents.

An alternative extraction procedure using sodium diethyl dithiocarbamate was given by Platte[300] for iron, copper, nickel, zinc and cadmium. Tenny mentioned[388] that samples with a high suspended solids content must be digested by heating with nitric acid for thirty minutes. Ion exchange and extraction were used by Chao et al.[84] to determine silver down to 0.1 μg/l, while gold was determined in natural waters by Zlatkis et al.[463] after concentration by chelation. In this method the sample is stood overnight with bromine-water and hydrochloric acid and then passed through a column containing a polyschiff base. The gold is eluted with hydrochloric acid and hydrogen peroxide. After oxidation with potassium permanganate, the chelated gold is extracted with MIBK and this solution aspirated in the instrument. Molybdenum was determined in lake-water after extraction of the oxine complex with MIBK.[85]

Mercury is an increasing hazard from insecticides and fungicides, and from the effluent of chemical plant and paper mills. The sensitive flameless method for mercury (see Appendix 1) can be used to determine this element to an absolute sensitivity of about 5 ng. A cold digestion should be performed (page 174) to ensure that mercury is liberated from methylmercury and similar compounds.

Seawater

Although a scatter correction is advisable, particularly at low wavelengths, the presence of $2\frac{1}{2}$–3% of sodium chloride and other salts does not appear to affect the response of most trace elements. It does, however, preclude the concentration of seawater samples by simple evaporation by an effective amount. Billings[58] found direct atomic absorption to be sensitive enough to investigate major ion ratios in seawater, and some alkali and alkaline earth metals have been determined against standards containing sodium chloride. Chelating resins were used[323] (see p. 100) to separate a number of metals from seawater. Undoubtedly APDC–MIBK extractions may also be used effectively in oceanographic analysis.

METALLURGICAL ANALYSIS

As atomic absorption is a technique for the determination of metals, it is not surprising that a very large number of its most valuable applications occur in the various metallurgical industries. Elwell and Gidley[1] pioneered atomic absorption methods for magnesium, zinc and lead as early as 1960. However, with the availability of hotter flames and a better knowledge of flame processes, a number of the interference effects reported and discussed between 1960 and 1965 are no longer of importance. In routine applications the technique frequently reduces to dissolution, dilution (usually with a spectroscopic buffer) and aspiration.

In metallurgical analysis, the need is always for the best accuracy and best speed. Control analysis is usually carried out under rigidly standardized conditions which have been designed so that the result is available in the shortest possible time. Check analyses are performed with the utmost care under less hurried conditions. Atomic absorption may be considered in both contexts. It will not, of course, compete in speed with the large direct-reading emission or x-ray spectrometers upon which most foundry control is now based. But smaller operators who cannot afford such equipment undoubtedly find a modified form of control with atomic absorption extremely valuable, particularly if a rapid method of dissolution can be evolved.

The potential accuracy and independent nature of atomic absorption analysis gives it an important place in check analysis and in the initial calibration of the working standards upon which the direct-reading spectrometers depend. That this is so is evidenced by the adoption of atomic absorption procedures by a number of national standardizing bodies.

Light alloys

After the work of Elwell and Gidley[1] and the determination of several more individual elements, the determination of copper, calcium, manganese and zinc in magnesium alloys and the same elements plus magnesium in aluminium alloys was described by Mansell et al.[248] Wilson[446, 448] used caustic soda dissolution and examined the capabilities of flames of three different orders of temperature in the determination of silver, copper, zinc, magnesium, zirconium and chromium, reporting that several interelement interferences were decreased in the hotter flames.

The analysis of aluminium alloys for nine elements after dissolution in hydrochloric acid was described by Bell,[52] and this has set the pattern for atomic absorption analysis of light alloys, dissolution in hydrochloric acid being much preferred to caustic soda for aluminium because of the lower final concentration of dissolved salts. A recommended procedure is set out below.

Other work on aluminium alloys has concerned the determination of particularly low traces of certain elements. Beryllium was determined without interference in the nitrous oxide acetylene flame by Peterson,[294] but Hirano et al[172] had to sublime the aluminium off as aluminium chloride in order to determine calcium traces when using an oxyhydrogen flame.

The following simple preparation procedure has been developed and will be found suitable for determining most elements in aluminium alloys:

Weigh out 0.5 g of sample, place in a 250 ml beaker and cover with a clock glass. Add 30 ml of hydrochloric acid (1 + 1). When the initial reaction has subsided, cool the beaker then add 3 ml of hydrogen peroxide (20 vol). Evaporate to 15 ml, cool and filter quantitatively into a 100 ml flask and make up to the mark. Standards should be made up to contain 0–25 mg/1 of copper, iron, magnesium, manganese, nickel, zinc and any other element in a 0.5% solution of high purity aluminium in 15% hydrochloric acid. Samples and standards for magnesium determination should be run with the nitrous oxide acetylene flame, or should contain 0.1% of strontium or lanthanum as chloride. The standard containing aluminium only gives the reference base line from which the other standards are measured. For concentrations higher than the ranges given, the sample solution is diluted with 0.5% aluminium solution and the final result multiplied by the dilution factor. Use the analysis lines and flames suggested in Appendix 1.

Calcium can only be determined in aluminium alloys using the nitrous oxide acetylene flame. The temperature of this flame eliminates most of the depressive interference of aluminium on calcium, but ionization occurs. Addition of an alkali salt, or dissolution of the sample in caustic soda followed by oxidation with nitric acid provides a solution with which flame ionization too is suppressed, and the detection limit of calcium is then about 0.001 % or less in the alloy.

If silicon is to be determined, a slightly different dissolution procedure is employed in order to ensure that the silica is dissolved and retained:[310]

To 0.5 g of sample add 10 ml of water and, in small portions, 8 ml of hydrochloric acid. After allowing the reaction to subside add 15 ml of hydrogen peroxide (50 vol) in small portions and heat. Cool, add 5 ml of hydrofluoric acid and dilute to 100 ml.

Polythene or PTFE ware must be used for this method. Standard solutions should contain 0.5% of high purity aluminium as well as a range of 0–250 mg/1 of silicon.

Similar methods may be used for the analysis of magnesium alloys, though in this case it may well be found unnecessary to add magnesium to the standard solutions.

Zinc alloys

Zinc base alloys also appear to present very few difficulties in routine analysis. The determination of a number of elements in both zinc and lead was detailed by Jimenez Seco[186] while zinc die-cast alloys were analysed by Smith et al.[366] using an internal standard method with a double channel instrument in order to achieve a high level of precision. In the latter method samples were prepared by dissolving in the minimum amount of hydrochloric acid, oxidizing with nitric acid, and then making up to a 1% solution. Tin, lead, iron, aluminium, magnesium, copper and cadmium were determined.

A recommended procedure for zinc alloy is:

Dissolve 2 g of the zinc alloy in 20 ml of 50% nitric acid, boil to expel nitrous fumes, transfer quantitatively to a 100 ml calibrated flask and make up with water. This solution is used for magnesium, iron and aluminium, and diluted by a factor of 2 for lead, cadmium and copper. Calibration standards are made by adding quantities of stock standards, of the concentration given in Table 8, to the amount of zinc solution (made by dissolving 100 g of pure zinc in 500 ml of nitric acid and making up to 1 litre) also indicated there, and making up to 100 ml with water.

TABLE 8

	Diluted stock solution (p.p.m.)	Range added (ml)	Zinc solution 100 mg/ml (ml)	Equivalent %
Lead	500	0–10	10	0–0.5
Iron	50	0–12	20	0–0.03
Magnesium	20	0–15	10	0–0.03
Aluminium	200	0–12	20	0–0.12
Cadmium	100	0–10	10	0–0.10
Copper	500	0–20	10	0–1.00

Copper-base alloys

Most minor and major elements in copper-base alloys can be determined by atomic absorption quite routinely, though few published reports have appeared. No interference effects were found by Gidley and Jones[153] or Elwell and Gidley[130] when determining zinc and lead with a town's gas flame, after removal of the silicon from the dissolved sample. Eisen,[128] also using a low temperature flame, reported interferences which are now readily overcome with higher flame temperatures.

A recent textbook on copper analysis[131] gives atomic absorption procedures for cobalt, lead, silver and zinc, and suggests that other elements could also be determined because the methods would be very similar. Cadmium, for example, could be done with the same flame conditions as the lead, but the same sample preparation would apply to most elements.

A few special methods have been suggested for individual elements, including tellurium and selenium,[370] low concentrations of lead,[111] and arsenic,[447] and an ion exchange method was described[260] using Dowex-1 (the sample being dissolved in hydrochloric acid) to separate tin, cadmium and zinc from copper.

A general scheme for analysis of copper alloys is made more difficult by the tendency of tin to precipitate as metastannic acid from nitric acid solutions and the insolubility of lead chloride at higher concentrations of lead in hydrochloric acid.

Two solvent/acid mixtures were investigated by Johns and Price,[189] viz., hydrochloric/nitric and orthophosphoric/nitric acids, each component acid being present at a concentration of 25%. Nearly every type of copper alloy proved to be soluble in the former, and after the solutions are diluted to a working concentration, the tin remains in solution for at least twenty-four hours. At considerable dilution, a slight turbidity may be observed after eight hours in those solutions where 5% of tin is present in the sample, but if the hydrochloric acid concentration is increased when the sample is further diluted, no precipitation should occur.

Phosphoric/nitric mixture dissolved all copper alloys except those with more than about 5% of tin. On dilution, the solutions were found to be stable for at least a week.

The hydrochloric/nitric mixture is recommended for routine analysis, as the need to retain the solutions for more than a few hours seldom arises, and this solvent can be used for the simultaneous analysis of materials of a much wider composition range.

The air acetylene flame should be used for iron, lead, manganese, nickel and zinc determinations. There appear to be no chemical interference effects, but minor matrix effects make it necessary for the acid and major element concentrations to be matched between samples and standards. Aluminium and tin are determined in the nitrous oxide acetylene flame. Copper enhances both of these elements slightly in this flame, behaving in large amounts as an ionization buffer, and this is another reason for maintaining the appropriate copper concentration in the standards.

The method details are as follows:[189]

> Weigh 1.000 g of sample into a 250 ml beaker and cover with a clock glass. Add 20 ml of solvent acid (hydrochloric acid 25%, nitric acid 25%) and warm gently on a hot-plate. After dissolution, cool and transfer to a 100 ml flask and make up to volume. Undissolved solid at this stage is almost certainly silica and can be allowed to settle. Make further dilutions if necessary with water, but if the samples contain tin, add, before making up to volume, 10 ml of 50% hydrochloric acid. Calibration solutions are made by adding the quantities of the stock solutions detailed in Table 9 to 20 ml of 5% pure copper solution (copper metal dissolved in the solvent acid) contained in 100 ml flasks. If a sample dilution has been made, use proportionally less of the copper solution in the standards, and, if a tin-containing sample has been diluted, add 10 ml of 50% hydrochloric acid to each calibration solution. To each tin calibration solution add the extra amount of hydrochloric acid 50% indicated in Table 9. The standard containing only copper gives the reference base-line from which all the other samples and standards are measured.

The instrumental conditions should be chosen, particularly for major components, to give the best noise-free response. These may not be the same as for

TABLE 9. Calibration solutions

Aluminium							
ml of 1000 p.p.m. Al	0	1.0	2.0	4.0	6.0	8.0	10.0
Std. concentration (p.p.m.)	0	10	20	40	60	80	100
% Al (no dilution)	0	0.1	0.2	0.4	0.6	0.8	1.0
Iron							
ml of 100 p.p.m. Fe	0	5.0	10.0	15.0	20.0	25.0	30.0
Std. concentration (p.p.m.)	0	5	10	15	20	25	30
% Fe (no dilution)	0	0.05	0.10	0.15	0.20	0.25	0.30
Lead and Nickel							
ml of 200 p.p.m. Pb or Ni	0	2.5	5.0	10·0	15.0	20.0	25.0
Std. concentration (p.p.m.)	0	5	10	20	30	40	50
% Pb or Ni (no dilution)	0	0.05	0.10	0.20	0.30	0.40	0.50
Manganese and Zinc							
ml of 100 p.p.m. Mn or Zn	0	1.0	2.0	4.0	6.0	8.0	10.0
Std. concentration (p.p.m.)	0	1	2	4	6	8	10
% Mn or Zn (no dilution)	0	0.01	0.02	0.04	0.06	0.08	0.10
Tin							
ml of 1000 p.p.m. Sn	0	2.0	4.0	8.0	12.0	16.0	20.0
Std. concentration (p.p.m.)	0	20	40	80	120	160	200
% Sn (no dilution)	0	0.2	0.4	0.8	1.20	1.60	2.00
ml HCl 50 %	10.0	9.2	8.4	6.8	5.2	3.6	2.0

highest sensitivity—particularly the flow rates for the nitrous oxide acetylene flame.

Good results can also be obtained with the nitric/phosphoric acid solvent, in circumstances where this may be preferred. Calibration must be made with this solvent, for, in addition to slightly lowered sensitivities, a greater 'blank' absorbance is observed in the air acetylene, though not in the nitrous oxide acetylene flame. Both effects are probably caused by the persistence of small agglomerates of undissociated copper pyrophosphate in the medium temperature flame.

Although the above recommended procedures give satisfactory results for a variety of different copper base compositions, the high acid content of the final solutions may cause corrosion with prolonged use in some types of spray chamber and burner systems. It is probable that the tin can be retained in solution at hydrochloric acid concentrations of 5% or less if all solutions, samples and standards, are made up to contain 1% of tartaric acid.

Lead-base alloys and solders

Lead-tin alloys have long caused difficulties in wet analysis, because of the tendency of tin to precipitate from nitric acid.

Lead metal with low tin content will usually dissolve completely in nitric acid. In order to determine traces of tin, Perry[293] dissolved 10 g of sample in 15 ml of nitric acid (sp.gr. 1.42) and then added 15 ml of 2% ammonium fluoride solution to complex the antimony which might be present. The solution was made up to

100 ml. With a multislot burner and air hydrogen flame, the limit of detection was 1–2 p.p.m. of tin in the original sample.

The usual solvent for lead-tin alloys, and particularly solders, is hydrobromic acid bromine mixture:

> Dissolve the sample in 3:1 hydrobromic acid:bromine. When dissolved, warm to remove the excess bromine. Dilute to known volume with 10% hydrobromic acid—with which all subsequent dilutions should be made.

Considerable success with the dissolution of lead-tin solders, prior to the determination of trace and minor elements by atomic absorption has been claimed by Hwang and Sandonato[182]. The samples were dissolved in a special mixture of fluoroboric and nitric acids, which produced a clear solution of the solders. The solvent acid solution used was 70% nitric acid, fluoroboric acid* and water in the proportions 3:2:5 by volume, and this was prepared afresh just before each analysis:

> Transfer 1.000 g of sample to a high density polythene 100 ml volumetric flask and add 50 ml of the solvent acid. Insert the stopper and place the flask in an ultrasonic cleaning bath for about 15 minutes.
>
> Direct heat should not be applied, neither should the flask be left in the bath too long otherwise in the generated heat tin will precipitate as metastannic acid. As soon as a clear solution is obtained (the ultrasonic bath expedites this by thorough mixing) more solvent acid is added to make up to volume.

The solutions can be aspirated in this form, and further dilutions, if necessary, should be made with the solvent acid, otherwise insoluble compounds are precipitated. Minimal interference effects were claimed in the analysis of aluminium, silver, arsenic, gold, bismuth, cadmium, copper, iron, nickel, antimony and zinc. The nitrous oxide acetylene flame was used for aluminium and arsenic, air acetylene for the remainder. Fluoroboric acid/hydrogen peroxide has also been used recently to dissolve cable-sheathing alloy.[467]

Iron and steel

All the metals for which iron and steel samples are analysed can be determined by atomic absorption. The non-metals, carbon, sulphur and phosphorus cannot, unfortunately, be determined in the present state of the art. A considerable literature on ferrous analysis has built up since 1961, from which it is clear that there are many suitable procedures for a number of the elements, but until now, no single procedure by which a majority of the elements may be determined together. The problems lie both in the dissolution and in the various interelemental interferences.

A comprehensive review[351] of the applications of atomic absorption spectrometry to the analysis of iron and steel summarizes the earlier work and also makes a valuable contribution to the determination of some trace elements.

It is clear that many common elements, particularly chromium, manganese, cobalt, nickel, copper and molybdenum, can be determined using an air acetylene flame after the sample has been dissolved in any one of several possible solvent acids. Thus Belcher and his co-workers[51,201–3] usually preferred sulphuric-

* To prepare fluoroboric acid: to 200 ml of 40% hydrofluoric acid at 10°C add, in small quantities, 75 g of boric acid. Allow to dissolve and store in a polypropylene bottle.

phosphoric acid mixture while others have used simply hydrochloric acid–nitric acid mixtures. The advantage of sulphuric–phosphoric mixture is that tungsten and other acid-hydrolysable elements are thereby retained in solution.

The outstanding problem encountered in the analysis of these elements is the considerable depressive effect of iron upon the sensitivity of chromium and molybdenum in the air acetylene flame. The effect has been overcome to a great extent by the addition of various spectroscopic buffers, thus enabling trace amounts of these elements to be determined with adequate sensitivity and accuracy. Quantitative accuracy at higher concentrations has been more difficult to achieve because of minor interferences which are not completely eliminated and which vary according to alloy composition. This is probably one of the reasons which until now has precluded the adoption of atomic absorption as a method for 'complete' analysis of ferrous alloys. This situation, however, appears to be resolved by the use of the nitrous oxide acetylene flame[390] and a recommended general scheme is given at the end of this section after the discussion on individual elements. Given a method which is free from chemical interference effects, the precision required in most metallurgical analysis is achieved by great attention to details of operating technique, as outlined in the section on determination of major components. In particular, the reading accuracy for this type of analysis should be improved by use of the 'difference' technique.

Atomic absorption may be used as a sensitive means for determining low traces of usual and unusual elements. Sometimes this is possible by a direct procedure, but the elements to be determined may have to be concentrated after removal of the matrix elements. Mercury cathode electrolysis may be preferred, but where the wanted element, e.g. cobalt, also plates out, the iron must be extracted into isobutyl acetate.

For the purposes of an atomic absorption procedure, the elements to be determined fall into three groups. Manganese, nickel, copper, cobalt, lead (and probably magnesium), are all determined in an air acetylene flame; chromium, molybdenum (in absence of tungsten), titanium, vanadium, tin and aluminium require the nitrous oxide acetylene flame. All of these are best determined in a final solution which contains perchloric acid only, and thus lend themselves readily to being determined together after a single sample dissolution[391]. The third group contains elements which require special dissolution procedures—e.g., niobium and silicon in the presence of hydrofluoric acid, and traces of lead, calcium and other elements where a matrix removal step may be necessary.

Because any general dissolution procedure for steel analysis is dictated by the requirements of chromium and molybdenum, these two elements are discussed first. The methods are detailed on p. 126.

Chromium and Molybdenum. The depressive effect of iron on the absorbance of chromium and molybdenum in the air acetylene flame, obviated to a large extent by the addition of ammonium chloride,[44, 279] has been discussed in the section on interferences. Ammonium chloride was included in a general scheme based on dissolution in aqua regia[309], which allows lower concentrations of both elements to be determined with the full sensitivity associated with an air acetylene flame. Small variations in response are liable to occur which made the method unreliable when extended to alloying concentrations of these elements. These variations are almost certainly due to a difference in oxidation state of the two elements between samples and standards. It is readily shown that a calibration curve obtained by using potassium dichromate (Cr VI) has a different slope from one using chromium metal dissolved in hydrochloric or even hydrochloric–nitric acid mixture (predominantly Cr III). Oxidation of Cr III to Cr VI in the presence of hydrochloric acid or high concentrations of chlorides must be carried out with care otherwise chromium is volatilized off as chromyl chloride.

Kirkbright et al.[208] found that the nitrous oxide acetylene flame overcame many interferences with the determination of molybdenum, and more recently, Ramakrishna[318] showed that iron in this flame had an 'enhancing' effect on molybdenum. Feldman et al.[138] reported the successful determination of chromium with a nitrous oxide acetylene flame, as also did Welz,[425] provided that calibration was effected with standard steels and not with synthetic standards having iron and nickel additions. The latter's findings again suggest a difference, probably in oxidation state of the chromium, between samples and synthetic standards.

Provided that standards are taken through an oxidation step with the samples, the same response is obtained.[390] The oxidation is carried out with perchloric acid, which, since it does not interfere with the chromium and molybdenum absorption response, forms the basis of the analysis solution. In perchloric acid solution, and with the nitrous oxide acetylene flame, the influence of iron on both chromium and molybdenum is one of slight 'enhancement'. This is probably a reduction in ionization as it increases with the amount of iron present until plateau values are obtained at not less than 0.7% of iron in the solution. In the method recommended for chromium and molybdenum, iron is always added, whatever the sample dilution, to maintain a fixed level of about 1%. In this way, chromium or molybdenum at any level can be determined. The method can even be extended to alloys other than steels.

The only limitation is when tungsten is also present in concentrations greater than 0.5%. This precipitates as tungstic acid when perchloric acid is the only solvent acid present, and some molybdenum coprecipitates giving low analytical results. Tungsten is retained in solution in the presence of phosphoric acid. This can be made the basis of an alternative procedure for molybdenum and tungsten, but is unsuitable for chromium as the latter again suffers interferences from the complex solution matrix.

Manganese. Many of the interference effects reported hitherto with various solvent acid mixtures were largely removed when perchloric acid was made the basis of the final solution[177] though, in an air acetylene flame interferences were experienced from high concentrations of cobalt, tungsten, chromium, nickel and molybdenum. These could be minimized by the addition of ethanol, which also helped to increase the sensitivity. Such interferences were not confirmed by Thomerson and Price[391] when the solutions prepared for chromium and molybdenum determinations were also used for determining manganese.

Nickel. Few workers have found chemical interferences in the determination of nickel. Most discussions have centred around the choice of acid solvent and the analytical line for measurement.

Sulphuric-phosphoric acid will retain tungsten in solution, but aqua regia is quicker. No chemical interferences are found in the presence of perchloric acid, and nickel is successfully determined in the solution prepared as recommended here for chromium and molybdenum.

Unless a monochromator of very good resolution is utilized, the nickel absorption line 232.0 nm may give a markedly curved calibration graph because of the breakthrough of non-absorbable radiation from a nickel line at 231.98 nm. The alternative line 341.5 nm is often measured instead.

Copper. No significant interferences have been reported whatever the acid solvent used or whatever major elements other than iron itself are present. Excellent results are obtained provided iron is incorporated in the standard. Copper can thus be determined by the method given.

For traces of copper, the extraction of its diethyldithiocarbamate into methyl isobutyl ketone was described by Atsuya.[40]

Cobalt. As with nickel, no significant interference effects have been found with cobalt in steels with any of the acid solvents. Iron increases the response of cobalt in perchloric acid solvent but the effect is overcome when the calibration standards contain the appropriate amount of iron as in the method given.

For determining traces of cobalt, Scholes[351] recommended the extraction of iron into isobutyl acetate.

Titanium. Titanium was not determined in steel until after the establishment of the nitrous oxide acetylene flame for routine use. Bowman and Willis[70] dissolved samples in aqua regia and fumed with sulphuric acid. The interference of iron depended upon both sulphuric acid concentration and flame conditions. Headridge and Hubbard[167] used hydrofluoric nitric acid mixture and finally made up the solution to contain 50% ethanol. The titanium response was increased and, with iron added to the calibration solutions, no interelemental interferences were found. Mostyn and Cunningham[280] found that the addition of potassium chloride to all solutions based on nitric hydrochloric acid solvent overcame all interferences, even that of iron, and so standards did not need to have iron added.

The same situation exists when the analysis solution is based on perchloric acid, although if mixed standards are made up according to the method given here, iron will automatically be present. Titanium can therefore be determined under the same conditions as the other elements treated in the scheme without loss of sensitivity.

Tin. No information about tin in steel has been published hitherto, probably because of its poor sensitivity, and the consequent limited value of atomic absorption as a direct trace method. In the nitrous oxide acetylene flame there is a small increase in sensitivity in the presence of iron, but no other detectable interferences. A 1% sample solution in perchloric acid enables tin to be determined to about 0.01%.

Vanadium. Vanadium has been determined in steel after dissolution in phosphoric sulphuric acid mixture[82] and with perchloric acid in addition.[213]

In a final solution based upon perchloric acid, iron enhances the vanadium absorption and iron must therefore be present in the standards. No other interelement interferences are experienced in the nitrous oxide acetylene flame.

Aluminium. Aluminium was determined in steel by Amos and Thomas[35] shortly after the nitrous oxide acetylene flame was first introduced. 'Soluble aluminium' was included in a scheme based on hydrochloric/nitric acid dissolution given by Price and Cooke[309]. An extraction method was described by Clarke and Cooke[89] in which the cupferron complex was extracted into methyl isobutyl ketone, giving 10- or 20-fold concentration increase, and a further sensitivity increase of 2 times by use of the organic solvent.

Konig et al.[217] also used hydrochloric–nitric acid mixture to dissolve steel samples, and insoluble residues etc. were fused with borax-soda and the enhancing effect of certain organic solvents was investigated.

In the perchloric acid-based analysis solution, and nitrous oxide acetylene flame, iron increased the sensitivity of aluminium, and was therefore added to the standards. No other interferences are found.

Lead. Since Elwell and Gidley[130] first mentioned the possibility of steel analysis. lead has been determined to 0.001% in steel by Dagnall et al.[111] by a double extraction method. A direct method allowed concentrations to 0.0025% to be determined[89,306] with a 2% sample solution and an air propane flame to improve lead sensitivity.

Again, in the perchloric-based solution, the presence of iron increases the sensitivity of lead, even with an air acetylene flame. No other interferences are found, and levels down to 0.002% can be determined.

Scheme for the analysis of steels

The following scheme has been found to be successful for all steels except those containing tungsten. A single weighing and preparation procedure is needed for eleven common elements.

Dissolve 1 g of sample in 10 ml of hydrochloric acid (sp. gr. 1.18) and 5 ml of nitric acid (sp. gr. 1.42). When the reaction has subsided add 10 ml of perchloric acid (sp. gr. 1.54). Evaporate slowly until the solution is fully oxidized and fumes of perchloric acid appear. (If the solution contains chromium it will probably turn red at this point.) Fume for five minutes, then cool and dissolve the soluble salts in 50 ml of water. Filter and dilute to 100 ml.

Prepare calibration standards by adding appropriate volumes of stock solutions, to

TABLE 10

	Basic % range	Wavelength nm	Flame type
Manganese	0–0.2	279.5	air acetylene stoic.
Nickel	0–0.4	232.0	air acetylene lean
Chromium	0–0.5	357.9	N_2O acetylene rich
Molybdenum	0–0.2	313.3	N_2O acetylene* stoic.
Copper	0–0.2	324.8	air acetylene lean
Vanadium	0–1.0	318.4	N_2O acetylene stoic.
Cobalt	0–0.5	240.7	air acetylene lean
Titanium	0–2.0	364.3	N_2O acetylene rich
Tin	0–0.5	224.0	N_2O acetylene stoic.
Aluminium	0–0.5	309.3	N_2O acetylene rich
Lead	0–0.2	217.0	air acetylene lean
Tungsten	0–5	255.1	N_2O acetylene stoic.

* The acetylene flow rate for molybdenum must be carefully adjusted to give maximum height of red feather with no luminescence.

cover the basic ranges (Table 10) to 1 g samples of pure iron (B.C.S. 260/3 is suitable). Then proceed as for the samples. Chromium and molybdenum stock solutions in particular must be added before the fuming. Mixed standards (i.e. one set of standards containing ranges of any or all of the elements to be determined) can be prepared, provided the iron content is maintained at 1 %.

Further standard ranges can be prepared to cover alloy concentrations up to 5 %, reducing sensitivity by one of the means described in Chapter 4 if necessary. To determine concentrations over 5 % the sample is diluted up to five times, and sufficient stock iron solution is added so that the final iron concentration is always 1 %.

Molybdenum in Presence of Tungsten. Tungsten is retained in solution in presence of phosphoric acid, even when perchloric acid is used in the final solution. The preparation of the samples is given below. Both molybdenum and tungsten have also been determined, using phosphoric–sulphuric–perchloric acid mixture, by Knight and Pyzyna.[213]

Solvent acid mixture: to 300 ml of water, add 100 ml of perchloric acid (sp. gr. 1.54) 100 ml of phosphoric acid (sp. gr. 1.75) and 100 ml of sulphuric acid (sp. gr. 1.84).

Dissolve 1 g of sample in 50 ml of the solvent acid mixture and heat gently. When dissolution is complete, oxidize with dropwise additions of nitric acid and evaporate until the first fumes of perchloric acid appear. Cool, dilute, filter and make up to 100 ml. Prepare calibration standards by adding appropriate quantities of molybdenum and tungsten stock solutions to 1 g of pure iron and take through the dissolution procedure outlined above.

Direct Determination of Other Elements in Steel. Magnesium is determined satisfactorily to about 0.001 % in steel by the simple procedure of Belcher and Bray[50] which has been used by many workers since. A 0.1 % solution of the sample in 1 % hydrochloric acid (or 0.5 % sample in 5 % hydrochloric acid) should contain about 0.5 % of strontium or lanthanum to overcome the effect of aluminium in the air acetylene flame. Other elements do not interfere.

Silicon is brought out of solution with the perchloric acid procedures described. Low concentrations of silicon in steel may be determined with the nitrous oxide acetylene flame using the following dissolution technique:

Dissolve 1 g of sample in 20 ml of 50 % hydrochloric acid and carefully add 15 ml of hydrogen peroxide (50 vol). Boil, cool and dilute to 100 ml. Standards should contain iron as well as a range of silicon concentrations added as sodium silicate solution.

Higher silicon levels, e.g. as in cast irons, may be determined (Price and Roos[310]):

Dissolve 1 g of sample in 10 ml of hydrochloric–nitric (4:1) acid mixture in a PTFE cylinder or flask. Cool, add 2 ml of hydrofluoric acid and dilute to 100 ml in a PTFE cylinder or flask. If standards are made up with sodium silicate solution, a little sodium chloride solution should be added to both samples and standards to equalize the ionization buffering effect.

Niobium is determined in steels down to about 0·05 % after a similar dissolution.[468]

Separation of the Matrix Elements. The following electrolysis procedure can be used to separate iron, nickel, chromium, cobalt, copper and tin:

Dissolve 2 g of sample in 20 ml of nitric acid—sulphuric acid—water (2 + 1 + 4) mixture. Take to dryness and fume for 5 minutes. Alternatively, after dissolution in aqua regia, fume with perchloric or sulphuric acid. Make up to 20 ml with water, then electrolyse using a mercury pool cathode and platinum anode, connected to a d.c. supply giving 15 A at 5–6 V.

Trace amounts of aluminium, calcium, niobium, magnesium, vanadium and zirconium can then be determined, simply by concentrating the remaining solution, or performing further extractions or separations should these be found necessary.

Calcium was determined by Scholes[351] to 0.002% in maraging steel. After removal of the iron by electrolysis the solution was made up to 100 ml. Strontium chloride was added to prevent interference from phosphate, vanadium and titanium in the air acetylene flame. Calcium can also be determined in the presence of iron using the nitrous oxide acetylene flame, with the same order of sensitivity.

Removal of the iron itself by solvent extraction depends on the solubility of ferric chloride, in strongly acid solutions, in a number of organic solvents. According to Morrison and Freiser[278] 99.9% of iron is removed from 7.75–8.0 M-hydrochloric acid solution by shaking for several minutes with an equal volume of isopropyl ether. Isobutyl acetate is also effective, and is used much in iron and steel laboratories, a pH of about 1 usually being recommended for the aqueous phase. Amyl acetate or methyl isobutyl ketone may also be used. Most elements common in steels remain in the aqueous phase though vanadium V, antimony V, gallium III and thallium III are also extracted.

The aqueous phase may be analysed direct for elements which are depressed in the presence of iron, or concentrated to improve detection limits of trace elements. Further extraction can be made using the ammonium pyrrolidine dithiocarbamate—methyl isobutyl ketone system. To determine very low traces of lead for example, Dagnall et al.[111] first extracted the iron into amyl acetate and then selectively extracted lead as iodo-plumbate into methyl isobutyl ketone.

Arsenic was determined in cast iron by Menis and Rains[268] after dissolution in nitric acid and extraction of the iron with 2 thenoyl-trifluoroacetone into carbon tetrachloride. Arsenic was measured in the remaining aqueous solution using an argon hydrogen diffusion flame (see Arsenic, Appendix 1) and an electrodeless discharge source (though the latter would have no particular advantage over a stable hollow cathode).

Tellurium was extracted into amyl acetate as the diethyl dithiocarbamate complex by Marcec et al.[249] to determine traces down to 5 p.p.m. in iron or steel. Samples are dissolved in concentrated mineral acids, and tellurium atomized in an air acetylene flame.

Ferro-alloys

In general, the atomic absorption method would be used for trace rather than major elements in ferro-alloys. Any convenient method of dissolution can be employed, provided the standards are made up to match the acidity and major components.

The following procedure may be used, for example, to determine aluminium and calcium in ferrosilicon, and could undoubtedly be extended to other elements:

Dissolve 1 g of sample in 10 ml of nitric acid and 10 ml of water in a PTFE beaker. Cover with a polyethylene lid and heat to boiling. Add 40% hydrofluoric acid dropwise through a hole in the lid until the initial reaction has subsided. Make further additions of hydrofluoric acid until all the silicon has been removed, washing down the sides of the beaker with water as necessary. Then evaporate to a volume of 5 ml and add 15 ml of 60% perchloric acid. Heat to fumes and continue fuming for 20 minutes. Cool the mixture, and add 50 ml of water, warming to dissolve. Cool again and make up to 100 ml in a calibrated flask. Aspirate this solution for the aluminium determination. For calcium, dilute 10 ml of this solution to 50 ml, adding in 4 ml of a solution containing 5% of iron as ferric chloride.

Standards should contain 100–500 mg/l of aluminium (equivalent to 1–5%) the presence of iron not being necessary, and 0–12 mg/l of calcium (equivalent to 0–0.6%) in 0.4% iron. The nitrous oxide acetylene flame is used for both elements.

Major components in ferro-alloys were determined by Smith et al.[364] The materials were dissolved, where possible, by a solution procedure similar to that given above, though a sodium peroxide fusion was required for the insoluble residue of ferrochromium. The elements determined were vanadium, titanium, silicon, niobium, boron, manganese and chromium, and the nitrous oxide acetylene flame was used throughout.

Nickel and high temperature alloys

Andrew and Nicholls[38] first analysed nickel for magnesium. Even with a low temperature flame, no interference was reported from aluminium and silicon. The analysis of nickel alloy was also described by Dyck[126] who dissolved the samples in hydrochloric acid and atomized in an air acetylene flame. Standards were prepared from pure carbonyl nickel. Various flame conditions were investigated.

In a textbook on the analysis of nickel,[236] procedures are given for the determination of traces of zinc and lead after dissolving 10 g of sample in 25 ml of 8 M-nitric acid and making up to 100 ml.

Iron was determined in high temperature alloys by Cunningham[98]. The samples were dissolved in aqua regia, but ammonium chloride was added to overcome the effects of nitric acid in an air acetylene flame. Cobalt and nickel did not interfere significantly. Welcher and Kriege analysed both nickel-base[423] and cobalt-base[424] high temperature alloys. Samples are dissolved in aqua regia (or, if they contain tungsten, niobium or tantalum, hydrofluoric–nitric acid mixture) and the nitrous oxide acetylene flame is used for all elements except iron. The tendency of cobalt to form a number of different complex ions may cause variations in response, and this may be overcome by evaporating both samples and standards twice to dryness with hydrochloric acid before finally making up to volume.

The noble metals

All the noble metals except osmium show good sensitivity, though numerous interference effects have been reported when these elements have been determined in presence of each other. Recent work on platinium and rhodium has indicated that many of these interferences are overcome in the very hot flames now available, though sensitivities are usually not improved. It is expected that investigations along these lines will soon be extended to show that better response is obtained from all the noble metals at higher temperatures.

A review[48] of atomic absorption and other spectrochemical methods for the determination of the noble metals covers the literature up to 1967.

Gold can usually be determined in ores and alloys if these dissolve in aqua regia. Interferences reported in early work using low temperature flames are much reduced in air acetylene which gives similar detection limits.

For determining low concentrations of gold several extraction methods are available. Gold can be extracted virtually quantitatively from 3 M-hydrochloric acid into methyl isobutyl ketone,[376] the distribution coefficient being greater than 1000. Samples digested in aqua regia and taken to dryness may be redissolved in hydrochloric acid, diluted and shaken with methyl isobutyl ketone[394,395] after the addition of a little hydrobromic acid.

Palladium is about as sensitive as gold, and with the air acetylene flame relatively few interferences have been reported. Ginzburg et al.[155] used a chemical procedure for concentration, precipitating with lead and copper sulphide, but in the presence of these and other elements no other preparation was necessary. Erinc and Magee[133] extracted palladium from solutions of platinum alloys as the thiocyanate–pyridine complex from hydrochloric acid solution into hexone achieving a detection limit of less than 1 p.p.m. in the extract.

To determine palladium in silver, Takeuchi[386] removed the silver by precipitation as chloride and then extracted the palladium as diethyl dithiocarbamate into methyl isobutyl ketone.

Rhodium is also quite sensitive, but appears to be much affected by interferences, both from other noble metals and from other elements and acid radicals. Strasheim and Wessels[375] described a complex pattern of depressions by copper, lithium and nickel and enhancements by zinc, magnesium, strontium and iron in a low temperature flame, and Deily[120] showed that in organic solvents, the solvent effects are minimized in lean flames, but that the burner position is critical. Johns and Price[190] showed that there are residual interferences in an air acetylene flame similar to those reported by Strasheim,[375] but that these are largely overcome in the presence of 1 % of sodium sulphate. Best results with sodium sulphate are obtained with a lean flame, but in air acetylene the calibrations tended to be non-linear. With a nitrous oxide acetylene flame and no buffer, no interferences were detected, the sensitivities were maximal as with sodium sulphate and the air acetylene flame, and the calibration curves were linear over a greater concentration range.

Platinum was also determined by Strasheim and Wessels[375] who found that, in contrast to their experience with rhodium, the interferences in the low temperature flame were much reduced in presence of copper. Pitts et al.[298,299] investigated the behaviour of platinum in both air acetylene and nitrous oxide acetylene flames, following a method by van Loon[409] in which the serious interferences of palladium, gold and silver on platinum in the air acetylene flame were eliminated by the addition of 1 % lanthanum chloride. Pitts et al. found that in the presence of lanthanum, with the air acetylene flame, most interferences from other noble metals, from base metals and from several common anions were virtually eliminated. The releasing action of lanthanum in this context was discussed. If the nitrous oxide acetylene flame is used, as has already been seen for rhodium, no interferences are detected, though it is stated that the sensitivity of platinum is lower than in air acetylene by a factor of about 4.

Silver is dissolved in nitric acid, but for the determination of other precious metals van Loon[409] used the following dissolution procedure:

> Add 5 ml of nitric acid to the sample, heating to leach the silver. Evaporate to 0.5 ml and add several ml of hydrochloric acid—continuing small additions until no further gases are evolved. Transfer to a volumetric flask. Add sufficient lanthanum chloride solution to make the final lanthanum concentration 1% and make up to the mark with 6 M-hydrochloric acid.

Silver itself is assayed in fine silver bullion[168] after dissolution in nitric acid by precipitating the bulk of the silver with standard sodium chloride solution, then determining the remaining silver in the supernatant by atomic absorption.

The 'new' metals

Zirconium and 'Zircalloys'. Two basic methods were given by Elwell and Wood[132] for dissolving zirconium alloys for atomic absorption analysis. For determining calcium and lithium, zirconium must be separated by ion exchange before the sample is aspirated.

> To 2 g of sample in 30 ml of water in a platinum dish, add 5.5 ml of hydrofluoric acid and allow to dissolve, then oxidize with a few drops of nitric acid. Add a solution of strontium chloride (≡6 mg Sr) and pass through an Amberlite IR:120H (20–50 mesh) column. Wash the column with four 30 ml portions of 1% hydrofluoric acid. Then elute with three 10 ml portions of 60% hydrochloric acid allowing the second of these to remain in contact with the resin for 20 minutes. The combined eluates are used for atomic absorption.

This procedure can be used for determining less than 10 p.p.m. of calcium in the alloy.

For copper and sodium, basically the same procedure is advised, though the ion exchange separation is omitted:

> Dissolve 2 g of sample in 4 ml of water in a platinum dish to which hydrofluoric acid is added dropwise till the dissolution is complete. Then add nitric acid dropwise until clear and make up to 10 ml in a polythene measuring cylinder and transfer back to the platinum dish.

These elements are all determinable in an air acetylene flame, and in view of the high concentration of the sample in the prepared solution in the second procedure, a multislot burner should be employed.

The other dissolution method involves fluoroboric acid and does not therefore require platinum ware.

> Dissolve 0.5 g of sample in 5 ml of 10% sulphuric acid and 2 ml of fluoroboric acid (see footnote p. 122). Clear with a few drops of nitric acid if necessary, cool and make up to 50 ml.

Although this method is given specifically for the determination of zinc, there would appear to be no reason why other elements should not be measured in the same solution. If a more concentrated solution is required, however, either the first or second methods may have to be used instead. The second method would be unsuitable if a nitrous oxide acetylene flame were to be used because of its high dissolved solids content.

Low traces of cadmium (0.1–1 p.p.m.) were extracted by Mizuno et al.[274] after dissolution in sulphuric, hydrofluoric, boric and nitric acids and adjusting to pH 8–10 with ammonia–ammonium citrate buffer, into chloroform as dithizonate. The cadmium was re-extracted back into hydrochloric acid for atomic absorption determination.

Titanium Alloys. Two similar dissolution methods based on fluoroboric acid were given by Elwell and Wood.[132] The first is recommended for determining magnesium, though presumably other trace elements could thereby be determined:

> Dissolve 0.5 g of sample in 15 ml of hydrochloric acid (30%) to which fluoroboric acid is added dropwise until the dissolution is complete. Make up to 25 ml.

For higher concentrations and alloying elements, specifically zinc:

> Dissolve 0.25 g of sample in 10 ml of 50% hydrochloric acid, adding fluoroboric acid dropwise until dissolution is complete. Make up to 250 ml.

Titanium is also dissolved by hydrofluoric–nitric acid mixtures, and the method for determining sodium in zirconium may also be used for titanium.

Niobium Alloys. The determination of copper and zinc in niobium is similar to the copper method given for zirconium; according to Elwell and Wood[132]:

> Dissolve 2 g of sample in 5 ml of hydrofluoric acid in a platinum dish. Add nitric acid dropwise until dissolution is complete, cool, make up to 10 ml in a polythene measuring cylinder and then return to the original platinum dish. Aspirate this solution for atomic absorption measurements.

Kirkbright et al.[206] determined copper in niobium and tantalum after extracting the 8-hydroxyquinoline complex from fluoride solution at pH 4.5 into ethyl acetate. Titanium and molybdenum in one-hundredfold excess interfered with the extraction, but were masked by the addition of hydrogen peroxide. Either air acetylene or air propane flames could be used.

Cobalt and zinc may be extracted into chloroform as thiocyanate diantipyridylmethane ion association complexes from solutions of niobium, tantalum, molybdenum and tungsten. The solutions should be adjusted to pH 3.25 in a citric acid medium and approximately 1.2 M in sodium thiocyanate.

Uranium Alloys. Magnesium was determined in uranium by Humphrey[179] without separation after dissolution (0.5 g sample) in 25 ml of hydrochloric acid. Hydrogen peroxide, if required to assist dissolution, is decomposed when the solution is evaporated to dryness and redissolved in hydrochloric acid. To avoid chemical interference effects, the method of standard additions was employed.

The bulk of the uranium matrix can be extracted into carbon tetrachloride with tributyl phosphate, allowing some concentration to be made[193] without affecting burner performance. Baudin et al.[46] described an analytical procedure for iron and aluminium (50–500 p.p.m. and 500–1000 p.p.m. respectively) based upon a sulphuric acid dissolution without separation. 1 g of the finely divided sample is attacked in the cold with 25 ml of sulphuric–nitric acid mixture (15 + 5, + 15 volumes of water). The solution is then evaporated to white fumes and the residue redissolved in water and made up to 100 ml, when it is about M/20 with respect to sulphuric acid. Iron is determined with an air acetylene flame and aluminium with nitrous oxide acetylene.

To determine antimony, iron and molybdenum in uranium and nickel, Walker et al.[414] dissolved the alloys in 8 M-hydrochloric acid and extracted these elements into *n*-amyl acetate. Samples of 25 g were used to attain a detection limit of 0.05 p.p.m in the alloy.

Complex alloys produced by nuclear fission have been analysed for molybdenum, ruthenium, palladium and rhodium by Scarborough[347]. The alloy is dissolved in hydrochloric–nitric acid mixture to which a little hydrofluoric acid was also added. An air acetylene flame was used for all the elements with a multi-slot burner. The absorbances given by mixed standards containing uranium were essentially the same as standards for individual elements with uranium added. Uranium thus eliminates the mutual interferences between these elements and the method may be used for a wide range of alloy compositions.

Electroplating solutions

With samples already in the form of solutions the control of electroplating baths is an ideal application for atomic absorption spectroscopy. In order to ensure an electrodeposit of the correct quality, concentrations of the major metals, metal-containing additives (used to provide characteristic plating properties) and trace impurities should all be checked. Impurities are introduced through bath to bath transfer of workpieces, in the commercial grades of the salts used to make the bath, in some superficial dissolution of the workpieces themselves, and in the continued use of hard waters.

If the electrodeposit is itself an alloy, its metal/metal ratios may have to be checked. Plate thickness can also be determined by atomic absorption spectroscopy.

When both major and trace elements are measured in one sample, the full concentration range may be as high as 50,000:1 and consequently a number of dilutions may be required. For the major elements a dilution of 5,000 is sometimes necessary. Alternatively the less sensitive absorption lines may be employed or the burner rotated.

Some plating solutions contain several major components, and therefore standards for trace elements should normally be made up in a matching matrix, unless it is proved that it does not interfere. Alternatively trace elements can be determined by the method of standard additions.

Plating solution analyses are not generally required to a very great degree of accuracy: impurities to $\pm 10\%$, additives to $\pm 5\%$ and main elements to $\pm 2\%$ is not unusual. On occasion it is even sufficient to report that the traces are below a certain level and that the major elements are above a prescribed minimum.

Copper plating solutions. Cyanide copper-plating baths contain 10–50 g/litre of copper as sulphate and 5–35 g/litre of sodium cyanide, while in a typical acid bath there may be 200 g/litre of copper sulphate in 3% sulphuric acid.

Iron, lead and zinc are the main impurities and unless they are present at very low concentrations (< 40 p.p.m.) good results are obtained with a 20 × dilution of the sample and comparison with simple aqueous standards. For very low levels where smaller dilutions must be used, the matrix compositions of standards and samples should be matched.

Nickel Plating Solutions. A typical nickel bath may contain, per litre: 300 g of nickel sulphate, 50 g of nickel chloride or sodium chloride, and 40 g of boric acid. Nickel itself is thus 50–60 g/litre, and copper, zinc, iron, lead, chromium, calcium and magnesium may all be in the p.p.m. range.

Shafto[355] used the standard addition method for some of these traces, while Whittington and Willis[433] determined copper and zinc in Watts-type nickel solutions, using a purified Watts solutions as a basis for standards. In the control of nickel plating baths, Parker[291] determined iron, lead, copper, calcium and magnesium in a 10 × dilution of the sample solution, comparing with simple aqueous standards. To determine zinc and chromium, standards had to be made up in the nickel-base solution because of enhancement of the former and depression of the latter in an air acetylene flame.

To determine the nickel itself, use of the alternative line, 352.5 nm, and rotation of the burner allows a dilution factor of 100 to be used instead of 5,000.

Chromium Plating Solutions. These are prepared by dissolving 200–500 g of chromic oxide per litre of water. The resulting solution is sufficiently acid to dissolve small amounts of copper, iron, zinc and nickel from the workpieces. The impurities are harmful only at the g/l level and the solutions can therefore be analysed after a 100 or 200 × dilution. All elements except iron (slightly depressed by high concentrations of chromium) can be determined by comparison with simple aqueous standards. Iron standards must contain the correct level of chromium. Chromium itself is not normally determined by atomic absorption in this case, as it is also required to know the concentration of Cr III (Parker[291]).

Zinc Plating Solutions. Alkaline zinc cyanide baths contain 30–55 g of zinc, 80–150 g of sodium cyanide and 20–35 g of sodium hydroxide per litre. Acid zinc baths contain 20–35 g of zinc, 20–25 g of sodium chloride and 12–15 g of boric acid per litre. Zinc itself may be determined in either solution by comparison with simple aqueous standards, using the alternative line 307.6 nm. Copper, lead, iron and tin can be allowed to build up to 0.1 g/litre without detriment, and when determined by atomic absorption, do not suffer interference from zinc in an air acetylene flame.

Cadmium Plating Solutions. Alkaline baths contain 20–25 g of cadmium, 80–120 g of sodium cyanide and 15–35 g of sodium hydroxide per litre. Copper in the range 1–10 p.p.m. and nickel (1–1000 p.p.m.) require control. Whittington and Willis[433] found that, while copper showed no interference, cyanide suppressed the response of nickel. The complex must thus be degraded with sulphuric–perchloric acid and standards must be made up from a 'pure' plating solution. Cadmium can be determined without interferences in an air acetylene flame after suitable dilution of samples.

Silver Plating Solutions. These usually contain the double potassium–silver cyanide, with silver 5–30 g, potassium cyanide up to 50 g per litre and small additions of potassium carbonate. Some chemists prefer to determine silver by a Volhard titration if great accuracy is required, however, silver does not exhibit interference effects in the air acetylene flame and simple standards may be used. Only copper among impurity elements is occasionally determined.

Gold Plating Solutions. Different types of gold bath usually have gold in the range 1–10 g/l as gold cyanide, with a small excess of potassium cyanide. Additions of cobalt and nickel cyanides may be made at the 1–2 g/l level. Copper, iron and zinc are allowed to accumulate up to 1 g/l, but chromium and lead are not tolerated above 10 mg/l.

Cobalt, copper, nickel, iron and zinc may be determined in an air acetylene flame with no interference effects. Chromium and lead show interferences and the standards must be based upon a 'pure' plating solution—though Kapetan[197] found also that lead was interference-free.

Although none of the trace impurities interferes with the determination of gold in the air acetylene flame, cyanide itself appears to give rise to differences in the response of gold. The effects are not overcome in the hotter nitrous oxide acetylene flame. Correct results are obtained, however, if the cyanide complexes are destroying by fuming with sulphuric–perchloric acid mixture and standards contain the same acids. Alternatively, standards may be prepared to contain the other major constituents in the correct concentration.

Rhodium Plating Solutions. A typical rhodium bath contains 2.5–5.0 g of rhodium as sulphate per litre of 5% sulphuric acid. Accumulations of up to 1 g/l of cobalt, iron and nickel may be tolerated in used baths. Over this amount they have a detrimental effect on the deposit. These elements can be determined without interference in the air acetylene flame after dilution of the samples.

Rhodium suffers fairly complex interference effects[190] which are largely overcome in an air acetylene flame in the presence of 1% of sodium sulphate. In the nitrous oxide acetylene flame the interferences are not evident, but because of ionization effects, the concentration of sodium (which is often present as a trace impurity in the bath) must be equalized in the standards. In fact it is much simpler to add an excess, e.g. 1%, of sodium sulphate to both samples and standards.

Analysis of Electrodeposited Films. Both composition and thickness of films can be determined by atomic absorption. Deposits may be removed from the base metal by selective stripping, mechanical removal or partial etch, and then dissolved in aqua regia. Kometani[216] applied the method to the determination of film thickness by assuming that the plate density is equal to the bulk density. Knowing the area from which the deposit has been removed, the thickness is calculated from the weight determined by atomic absorption spectrometry. The bulk density of gold for example is 19.3 g/cm^3 and the thickness is calculated from the formula:

$$\text{thickness} = \frac{\text{weight of deposit } (\mu\text{g})}{\text{area } (\text{cm}^2) \times \text{density } (\text{g/cm}^3) \times 100} \mu\text{m}$$

The majority of electrodeposited films may be examined by the above principles, simply by removing the film with mineral acid attack.

Deposited alloy compositions may also be determined. The copper–zinc ratio of brass which has been coated on to iron wire is arrived at with adequate accuracy after stripping the sample with ammonium persulphate and dilute ammonia solution. It is important to acidify the sample solution with hydrochloric acid before aspiration and to compare with standards containing the same major ingredients.

The analysis of both evaporated and electrodeposited films was also discussed by Woolley[454], who examined combinations of metals such as nickel–chromium, nickel–cobalt, nickel–iron as well as single element films of chromium, gold and aluminium.

Ferrites

The problems of ferrite analysis are very similar to those of alloys once the material has been brought into solution. This can usually be achieved using a pressure technique such as will be described in more detail under silicate analysis. Alternatively sealed tubes can be used:

> Weigh 50 mg samples of the finely divided ferrite material into strong Pyrex test tubes. Add 5 ml of 6M-hydrochloric acid and seal off the mouth of the tube. Heat to 210° C until dissolution is complete. Cool, open, and make up to 100 ml with water. Dilute further as necessary for individual elements and compare with simple aqueous standards.

Little interference between components should be experienced after concentrations have been brought within the working range by dilution.

INORGANIC ANALYSIS

This section includes analyses in the field of pure chemicals, ores and other mining applications, and siliceous materials, whether natural or man-originated. It should be recognized that this can at best be a guide to some of the ways of approaching the many types of analysis which are now possible.

Pure chemicals

Problems here are often identical to those encountered with metals—the analysis of metal salts for example. Marshall and West[251] discussed the particular problems in the determination of small traces of calcium, magnesium, iron and nickel in aluminium salts. Calcium and magnesium were coprecipitated with ferric hydroxide, and after redissolution were extracted with 8-hydroxyquinoline into methyl isobutyl ketone at pH 11. Similarly, iron and nickel were coprecipitated on hydrated manganese oxide, and then 8-hydroxyquinolinates extracted at pH 4.5. Chromium was extracted directly as Cr VI into methyl isobutyl ketone.[252]

The rapid analysis of zinc oxide for copper, lead, iron, manganese, nickel and calcium was described by Spitzer.[369] The samples were dissolved in boiling hydrochloric acid with addition of perchloric acid. As found with zinc alloys, zinc itself interferes only to a minimal extent.

A different approach was used by Vink[413] who separated the manganese from a solution of manganese dioxide in hydrochloric acid with added isopropyl alcohol on Dowex AG 1–X8 anion exchange column. Mutual interference between the impurities—cobalt, copper, iron, nickel and lead—was negligible.

Impurities in powdered elemental boron were determined by both atomic absorption and flame emission by Bedrosian and Lerner[49]. Powdered electrolytic and some types of hot-wire boron can generally be dissolved in nitric acid or sulphuric nitric mixtures. If the sample contains tantalum or tungsten, a combined acid attack, followed by evaporation with hydrofluoric acid, must be used, and the insoluble residue is then fused with sodium carbonate and taken up in hydrochloric acid to give a second solution. The simpler method yields better precisions for all elements and is used wherever possible.

High purity sodium metal was analysed by Scarborough et al.[348] The sample was dissolved with argon saturated in water vapour and then neutralized with hydrochloric acid. The wanted impurities are separated by the simple expedient of co-precipitating with lanthanum as hydroxides by bubbling ammonia, diluted with argon, through the solution. The precipitate is then redissolved in hydrochloric acid.

The impurities in uranium and uranyl compounds were determined by Walker et al.[414] after dissolution in 6M-nitric acid and extraction of the uranium into carbon tetrachloride with tributyl phosphate. The aqueous phase was then evaporated to dryness, taken up in hydrochloric acid and analysed for aluminium, calcium, cadmium, cobalt, chromium, copper, iron, magnesium, manganese, nickel, lead and zinc. The contamination of plutonium salt solutions from stainless steel apparatus was measured by Ganivet and Benhamou[150] who determined the iron, chromium and nickel at levels of 0.1–8 μg/ml in presence of 100 g/l of plutonium.

There are very many cases where methods are so simple—e.g. dissolution in an appropriate solvent—that they have not been reported. Atomic absorption procedures are now regarded as indispensable by manufacturers both of industrial chemicals and, particularly, of high grade laboratory reagents.

Mining and geochemical applications

It is often difficult to differentiate between the problems encountered in these fields of application. Only the uses to which the results are put (and sometimes the degree of accuracy required) are different. Geochemical analyses, including mining applications, have been the subject of a book[3] and methods of attack for many types of geochemical sample are given there.

For our purpose, however, materials in this category may conveniently be divided into siliceous and non-siliceous types, for the latter usually incur greater problems in preparation. In this section essentially non-siliceous materials are discussed, as silicate materials in general are treated as a separate subject in the section which follows.

Since Strasheim et al.[374] determined copper in ores in 1960, atomic absorption has been used to determine both major and minor wanted elements, as well as trace impurities in many different types of ore. Ores to be assayed for copper were leached by these workers in a hot mixture of hydrofluoric and nitric acids. This was said to be more successful than either sulphuric–nitric or hydrochloric–nitric mixtures, and other workers have used hydrochloric–nitric acid[134] and perchloric–nitric acid[362] when copper, zinc and lead were to be measured.

Vanadium was determined in copper ores after dissolution in sulphuric or hydrochloric acid solution, though for low vanadium contents, the iron had first to be extracted from 8M-hydrochloric acid solution into isopropyl ether. The hydrochloric acid itself interfered and had to be removed by evaporation[156] and aluminium was added to the prepared sample to improve the sensitivity. The same worker[157] also determined tellurium in copper ore after precipitation with stannous chloride in hydrochloric acid.

Lead ores may also be dissolved in hydrochloric–nitric mixtures[47] but the determination of tin in cassiterite requires a more difficult separation[169]. The ore must be ground with ammonium iodide and heated to 500°C. The tin iodide which volatilizes may be dissolved in hydrochloric acid. Addition of ammonium iodide to the final solution increases the sensitivity of the tin.

Molybdenum has been determined in siliceous ores[261] and in slags[302]. The molybdenum is leached with hydrochloric acid, and the residue digested with nitric, hydrochloric and perchloric acids, fumed and taken up in hydrochloric acid. Both workers used aluminium chloride to overcome the effects of interfering elements on molybdenum. The releasing action of aluminium chloride rises to a plateau at about 1000 p.p.m. of aluminium.

The presence of mineral deposits may be discovered through the analysis of plants for metallic elements. This form of biogeochemical prospecting was described by Hornbrook[176] who determined molybdenum in soils and plants—the latter after an ashing procedure similar to those discussed in the biological section.

Gold may be assayed in ores by methods similar to those mentioned on page 130. In most cases an attack on the ore material with aqua regia is satisfactory and the gold may then be concentrated by one of several methods. Law and Green extracted the gold directly into methyl isobutyl ketone,[231] having studied the conditions giving a clean extract in the presence of emulsion-forming insoluble residues. Reeves[321] used *p*-dimethylamino-benzal rhodanine to extract the gold from the aqua regia dissolution into amyl acetate, after taking to dryness and redissolving in hydrochloric acid. The absolute sensitivity of the method is 0.2 μg of gold and no interferences were detected from any metal likely to be present. Gold was precipitated, with tellurium, in the reduction method given by Toyoguchi[399] and methanol was added up to 10% to improve sensitivity.

The determination of platinum in basic rocks was described by Simonson[359]. After digestion with nitric–hydrofluoric acid and evaporation, the residue is redissolved in hydrochloric acid and extracted with dithizone into methyl isobutyl ketone. At this stage, only otherwise interfering elements are extracted. Platinum itself is not extracted until reduced to Pt II. The aqueous phase is therefore reduced with stannous chloride and again extracted into methyl isobutyl ketone. This platinum-containing extract is then aspirated into a lean air acetylene flame. A simpler extraction was described by Swider[383] and is a modification of that given earlier by Tindall[394]. After digestion first with aqua regia and then with hydrochloric acid, the samples were made up to volume and filtered. (Many base metals not present as silicates may be determined in this solution.) To an aliquot is added potassium iodide solution and the platinum is extracted into methyl isobutyl ketone. After further addition of the potassium iodide solution and two periods of shaking, the organic solution was aspirated into a multislot burner supporting a lean air acetylene flame.

Rhodium was determined in chromite concentrates after precipitation with tellurium or separation into the gold bead resulting from fire assay[350] although the latter results in incomplete recovery of the rhodium.

Soils, rocks and ores were analysed for bismuth[181] after digestion with nitric acid, and filtration. An air acetylene flame was used and standards were simply made up in nitric acid, few metals other than calcium giving detectable interference.

Methods for the atomic absorption determination of potassium, rubidium and strontium for geochronological purposes were given by Gamot et al.[149] Caesium and lanthanum were used together as a buffer. Accurate results were obtained for potassium by operating in a purified atmosphere. Strontium[173] and rare earths[200] have been determined in phosphate rocks. Strontium was extracted from the sample with hydrochloric acid, and phosphate removed by passage of the resulting solution through a cation exchange column, the cations being subsequently eluted with hydrochloric acid. Many rare earth elements, with the exception of cerium, are determined in a hydrochloric acid solution aspirated into a nitrous oxide acetylene flame, sodium chloride being added as ionization buffer. Main interferences were from silicate, fluoride and aluminium. Samples were prepared by fusing with sodium peroxide, followed by dissolution of the alkali-insoluble residue in hydrochloric acid. The rare earths are then precipitated with oxalic acid, ignited and redissolved in hydrochloric acid.

Some of the more sensitive (and less common) rare earth elements were determined in various rare earth minerals by van Loon et al.[410] Zirconium rare earth silicates were decomposed with hydrofluoric acid, while the calcium mineral was subjected to decomposition by alkaline carbonate fusion. The wanted elements were precipitated as hydroxides, dissolved in nitric acid and made up to a known concentration with lanthanum buffer. Solution detection limits ranged from 0.1 p.p.m. for ytterbium to 10 p.p.m. for lutecium.

There are many other examples of the application of atomic absorption analysis where sample dissolution presents very few problems, as in zinc, manganese and iron ores, limestones etc.

Siliceous materials

A number of types of material contain silica or silicates as one of the major constituents. Often the other major elements include calcium, magnesium, aluminium and perhaps iron or manganese, or sodium and potassium. Among such materials are silicate rocks, lavas and minerals, slags and refractories, ceramics, glasses, cements, ashes etc.

From the point of view of atomic absorption they have several problems and features in common. They require special methods to bring them into aqueous solution, but when in solution, all constituents (with the exception of phosphate if present) can readily be measured. There are, however, a number of interference effects to be overcome before accurate determinations of either major or minor constituents can be made.

An examination of the reported methods for the atomic absorption analysis of siliceous materials reveals that they fall into two distinct groups: those in which the sample is brought into solution by way of acid attack, and those in which the sample is fused with an alkali fusion salt or mixture. A further subdivision of both types of method is into those where the silicon is retained in solution and determined by atomic absorption and those where it is removed, either completely by volatilization, or precipitated as silica, and perhaps determined by gravimetry.

The main disadvantage of fusion methods is that they result in a prepared solution of high dissolved-solid content, which can usually be tolerated in the determination of major constituents where a greater dilution factor is applied, but which generally requires separation before minor or trace elements can be determined. On the other hand, with fusion methods, it may be easier to ensure that all silicon is present in solution.

Fusion Methods. By far the most common fusion medium is lithium metaborate, which may be used to attack silicate rocks, slags, cement and coal ash. Some workers substitute a mixture of lithium carbonate and boric acid, or prepare lithium metaborate from these beforehand.

Typical procedures have been given by van Loon and Parissis[411] and by Boar and Ingram[63]. The rock or ash sample, 0.2 or 0.5 g, ground to at least B.S.S. 250 mesh, is mixed with 2 g of lithium metaborate in a platinum or graphite crucible, and heated to 900°C in a muffle furnace for fifteen minutes or until a clear melt is obtained. If there is more than 15 % of iron as magnetite, 50 mg of ammonium vanadate is added to flux the iron. The melt is then cooled and placed in a beaker containing 8 ml of nitric acid and 150 ml of water. The mixture is stirred continuously until dissolution is complete. If the dissolution is not allowed to take place at room temperature the silica may form polymeric hydrated silicic acid which does not dissolve.

Aliquots are taken and diluted as necessary. Silicon, aluminium and titanium must be determined with the nitrous oxide acetylene flame, but no spectroscopic buffer is added for these elements—or for calcium and magnesium if these are measured in the same flame. Calcium and magnesium determined in an air acetylene flame require the addition of 1 % of lanthanum to the final solution to overcome interference by silicon, aluminium, borate and titanium. It is necessary to match the lithium, borate and acid concentrations of prepared samples and standard solutions in order to equalize degrees of ionization of the readily ionized elements in either flame.

In an alternative approach for routine analysis of slags,[319] the sample is fused with sodium carbonate and leached with hydrochloric–nitric acid mixture. The precipitated silica is filtered off. Aluminium, calcium, magnesium, manganese and iron were determined in iron, copper and lead slags, and silicon was determined in a separate sample.

Acid Attack. Most solvent acid mixtures contain hydrofluoric acid, but, depending on the way in which it is used and the other component of the acid mixture, it is possible to remove the silicon completely by vaporization as silicon tetrafluoride, or to keep the silicon in solution for determination by atomic absorption. The choice can therefore be made at the outset as to whether the silicon is to be determined with the other elements or perhaps gravimetrically in a separate sample. If the silica content is not more than about 25 % and 1 % accuracy is acceptable, then it is worthwhile determining it by atomic absorption.

Langmyhr and his co-workers[221, 230] found that hydrofluoric acid is a suitable reagent for the dissolution of siliceous material and showed that no silicon is lost by volatilization provided that an excess of the acid is used and that the solutions are not too strongly heated. This principle was used in the later work by Langmyhr and Paus[222] who distinguished between two possible groups of materials. The first includes those which can be decomposed easily below 112°C, the boiling point of the azeotropic mixture of water and hydrofluoric acid, using relatively simple equipment. The second group comprises materials which are more difficult to decompose, requiring to be attacked at higher temperatures inside a form of 'bomb'.

Three possible methods of attack can in fact be employed, depending on the material being analysed. For decompositions which take place at room temperature or temperatures up to 30–40°C, plastic beakers or platinum dishes, with lids to prevent loss by spray action, can be used provided that the volume of solution is not allowed to reduce appreciably. Where decomposition must take place at higher temperatures, e.g. on a boiling-water bath, plastic bottles or Erlenmeyer flasks with screw stoppers are employed. If still higher temperatures are necessary the 'bomb' technique is required. Examples are given of all three methods.

The principle of the retention of silicon in hydrofluoric-based solutions in open vessels at lower temperatures was also confirmed by Price and Roos[310] who used hydrofluoric acid to prevent the precipitation of silica in the atomic absorption analysis of some metals and alloys as well as cement. The dissolution itself must be carried out in either a platinum or a PTFE vessel. Subsequent dilutions can be made either in plastic volumetric ware which is now available, or the excess hydrofluoric acid may be complexed with boric acid allowing glass volumetric ware to be used.

A suitable procedure[330] for preparing cements for atomic absorption analysis is as follows:

> Weigh 0.500 g of powdered sample into a 100 ml polythene or PTFE beaker. Wash down the sides of the beaker and add 10 ml of hydrochloric acid (sp. gr. 1.18). Break up gritty particles with the end of a stirring rod. When the sample has dissolved (apart from remaining or precipitated silica) rinse down and remove the rod, then add 1.0 ml of hydrofluoric acid (40%). Swirl until all silica is dissolved, add 50 ml of boric acid solution (4%) and mix. Transfer to a 200 ml calibrated flask add 20 ml of lanthanum chloride solution (5%) and dilute to the mark with water. This solution is used for the determination of aluminium, iron, manganese, silicon, sodium, strontium and zinc.
>
> Transfer 10 ml of this solution to a 100 ml calibrated flask, add 5 ml of hydrochloric acid and 9.0 ml of lanthanum chloride solution (5%) and dilute to the mark with water. This solution is suitable for determining calcium, magnesium and potassium.

Calibration ranges for standards and brief operating conditions are summarized in Table 11.

A method similar to the above was also given by Langmyhr and Paus.[225]

Decomposition on a boiling-water bath may be necessary for certain kinds of cement, industrial silica, glasses, quartzite, sand, sandstone, coal ash, feldspars and steelmaking slags and sinters. The decompositions are most conveniently carried out in polythene or polypropylene bottles with screw stoppers.

> Weigh out 0.2 g of finely divided sample into the bottle, moisten with 1 ml of water, add 5 ml of hydrofluoric acid (40%), stopper the bottle, and place on a boiling water

bath.[222, 223, 227] If a clear solution is obtained add 50 ml of 2% boric acid solution, mix and transfer quantitatively to a calibrated flask. Make up to volume after adding an appropriate quantity of lanthanum chloride solution or other buffer required.

A useful variation of this technique was given by Reid et al.[322] for the analysis of steelmaking slags. For blast furnace slags, hydrofluoric acid alone is satisfactory, but sinters, O.H. and B.O.S. slags need hydrochloric acid as well. The higher pressure caused by this acid may be accommodated by placing the bottles in a

TABLE 11

Element	Wave-length	Flow rates (l/min) Acetylene	Oxidant		Compositon of standards Conc. range: p.p.m.	Boric acid %	Lanthanum %
Al	309.3	4	N_2O	5	0–125	1.0	0.5
Ca	422.7	1.4*	air	5	0–200	0.1	0.5
Fe	248.3	1.4	air	5	0–125	1.0	0.5
Mg	285.2	1.4*	air	5	0– 12.5	0.1	0.5
Mn	279.5	1.4	air	5	0– 2.5	1.0	0.5
K	766.5	1.0	air	5	0– 2.5	0.1	0.5
Si	251.6	4.3	N_2O	5	0–750	1.0	0.5
Na	589.0	1.0	air	5	0– 12.5	1.0	0.5
Sr	460.7	4.0	N_2O	5	0– 12.5	1.0	0.5
Zn	213.9	1.0	air	5	0– 0.5	1.0	0.5

* Sensitivity is reduced by rotating burner or using an emission (short path length) head.

domestic pressure cooker, which at its 15 p.s.i. (1.05 bar) setting gives a temperature of about 121°C. The weighed sample is transferred to the bottle, and the solvent acid added. For sinters this is 5 ml of 9 + 1 hydrofluoric hydrochloric acid and O.H. and B.O.S. slags, 5 ml of a 7 + 3 mixture of the same acids. After a suitable time in the pressure cooker, which may vary from $2\frac{1}{2}$ minutes for B.F. slag to 20 minutes for sinters, the pressure is released, the bottles are rapidly cooled, and an excess (e.g. 100 ml of 1 %) boric acid is added. As no solution is lost by evaporation, all additions may be made by pipette, and time is not wasted in making up to volume finally in a calibrated flask. In Reid's method the nitrous oxide acetylene flame was used for all elements (calcium, magnesium, silicon, manganese and aluminium) except iron, and sodium chloride was the only buffer used, to reduce ionization of calcium and magnesium.

If the decomposition results in the formation of a precipitate, the decomposition vessel must be closed again after the addition of the boric acid and heated again as before until a clear solution is obtained.

Materials which may require attack at higher temperatures may be treated in a specially designed pressure vessel or bomb. Typically, such a device would be made from aluminium and lined with PTFE.[324, 184] The 'Uni-seal' decomposition vessel operating at 110°C, described by Bernas[56], is available commercially. Such samples may also require the presence of other acids for complete dissolution.

Bauxite was analysed for silicon, aluminium, iron, calcium, titanium and trace constituents[224] by the general method outlined above, but after decomposition in a bomb for 30 minutes. Ferrosilicon required the addition of 1 ml of concentrated nitric acid[226] and some iron ores and slags required the addition of 1 ml of aqua regia.[228] Sulphide minerals and ores,[229] 0.2 g samples, were decomposed with 2.5 ml of hydrofluoric acid and 2 ml each of nitric and hydrochloric acid. The temperature for all these decompositions was 110° ± 5° C. A Uni-seal bomb was operated at 135°C by Sheridan[356] who digested samples of mineral slag and wool for 40 minutes with 48 % hydrofluoric acid plus a little aqua regia, prior to determining silicon, aluminium, magnesium and calcium. Before transferring the digest from the bomb, the excess of hydrofluoric acid was complexed with boric acid.

If a residue of undecomposed material remains in the bomb, it may be necessary to repeat the operation, the sample having been ground to a finer powder, with a prolonged heating time at maximum temperature.

It is always essential that the excess hydrofluoric acid should be complexed by addition of boric acid as soon as possible after the decomposition vessel or bomb has been cooled and opened. This allows subsequent handling to be done in glassware, and also prevents precipitation of lanthanum fluoride if lanthanum is subsequently added as a spectroscopic buffer.

The addition of spectroscopic buffers depends upon the composition of the sample, the elements being determined, and the flames to be used. To some extent boric acid behaves as a releasing agent, but the releasing action of lanthanum, combined with its ionization buffering effect at higher temperatures makes it a useful addition, whatever flames are employed.

The elements determined by Langmyhr and Paus after hydrofluoric acid decomposition are silicon, aluminium, iron, magnesium, calcium, sodium, potassium, manganese and titanium. Chromium, copper, nickel, cobalt, lead, tin, zinc and vanadium may also be determined in slags, ores and minerals as required.

The interferences reported by most workers in this field follow a similar pattern. In the nitrous oxide acetylene flame the response of silicon is somewhat enhanced in the presence of calcium, iron and sodium but this must be an ionization effect and it is constant in the presence of lanthanum. Phosphate depresses silicon slightly but not in the presence of lanthanum. Aluminium is depressed by silicon and calcium when these are present together and again lanthanum removes the effect. Both aluminium and titanium sensitivities are greater in presence of fluoroboric acid, and it is essential that acid concentrations in samples and standards be matched. Magnesium shows no interferences in the nitrous oxide acetylene flame, and calcium alone is depressed because of ionization, though this is overcome by addition of potassium, another alkali metal, or at this temperature even by lanthanum.

In the air acetylene flame, depression of magnesium and calcium by silicon, aluminium and phosphorus is overcome by lanthanum and ionization of the common alkali metals is minimized by addition of one of the less common alkali metals (lithium or caesium is used). Iron, manganese and copper are not interfered with in air acetylene, and the depressive effect of iron on chromium is not encountered in nitrous oxide acetylene.

Removal of Silicon. The foregoing methods were evolved with the definite object of retaining silicon quantitatively in solution, but if this element is not to be determined for any reason within a general scheme, the methods of attack become somewhat simpler. The silicon is then either volatilized off completely by fuming with hydrofluoric acid and sulphuric or perchloric acid, or is dehydrated with either of these acids and separated by filtration.

This former principle was applied to the analysis of glass by Adam and Passmore[29]. A sample weight of 0.1 g was moistened with 2–10 ml of water and 1 ml of perchloric acid and 5–20 ml of hydrofluoric acid were added. The mixture was digested on a steam bath and then fumed to near dryness. After a second fuming with 1 ml of perchloric acid the residue was dissoved in perchloric acid and made up to volume. A spectroscopic buffer was added to the aliquot used for determining magnesium, calcium, strontium and barium. Both Belt[53] and Thompson[392] analysed silicate rocks by this type of procedure, and performed a statistical analysis of results obtained on two standard rocks. Belt's procedure was to attack 0.5 g of sample with 2 ml of water, 20 ml of hydrofluoric acid and 10 ml of nitric acid. After standing for two hours, 2 ml of perchloric acid were added and the mixture fumed until no more fumes were evolved. The residue was then dissolved in hydrochloric acid and made up to volume. Lanthanum was added to overcome the effect of aluminium on magnesium and calcium in an air acetylene flame, but no buffer was required for sodium, potassium, manganese or iron. Thompson's procedure was similar except that iron was determined in the solution to which lanthanum had been added. Acid and buffer concentrations in samples and standards were carefully matched. Coefficients of variation quoted by both authors for a common sample NBS W–1 are remarkably similar, though Belt's reproducibility figures were obtained on one day, while Thompson's were obtained from repeated recalibrations over a period of some months (see Table 12).

TABLE 12

	Content % (as metal)	Coeff. of variation %	
		Belt[53]	Thompson[392]
Calcium	8.1	1.6	1.9
Magnesium	4.12	1.9	1.1
Iron	7.6	0.7	2.1
Sodium	1.57	1.9	2.5
Potassium	0.50	5.0	5.2
Manganese	0.136	1.0	1.2

ORGANIC MATERIALS

The determination of trace and major concentrations of metals in organic matrices are often some of the easiest analyses to carry out by atomic absorption. In this section we discuss types of sample that are not soluble in aqueous media but which may form solutions with other organic liquids, samples from which it may be necessary to remove the organic matrix by oxidation, as well as those which are simply soluble in aqueous media. Biochemical samples, too, could be classed in this latter general category, but for convenience are discussed separately.

As organic solvents usually increase the sensitivity of many metals, due to the more highly reducing nature of the resulting flame, and perhaps to the greater rate of vaporization of the solvent, direct nebulization of an organic solution usually results in improved sensitivities. There may be a problem in providing non-aqueous standards which are known to give the same response in the flame as the sample itself.

Apart from such direct methods, the wanted metals can sometimes be extracted from the organic matrix with mineral acids, the matrix itself can be removed by wet or dry oxidation, or small samples can be effectively attacked by the oxygen flask or bomb method.[256]

Among the many types of non-aqueous samples analysed by direct procedures are petroleum and petroleum products, materials encountered in the paint, oil and colour industry, plastics, fibres etc.

Petroleum products

Crude Oils and Feedstocks. The determination of trace metals in crude oils is necessary before such oils are refined and submitted to catalytic 'cracking', because some metals, in particular nickel, copper, iron, sodium and vanadium, may poison the catalyst and reduce the efficiency of the plant. These metals are often present, at least in part, as volatile organometallic compounds which accompany the hydrocarbons during fractionation.

The sources of oils, too, may often be identified by the relative concentrations of some trace elements, and rapid determination is therefore useful not only to oil companies but also those engaged in forensic and environmental pollution investigations.

Since Robinson first determined sodium,[326] this method has been extended to nickel, copper and iron by Barras and Helwig[45] and to nickel by Trent and Slavin[400] and by Kerber.[199]

Barras diluted gas oil feedstocks 1 + 2 with *n*-heptane, comparing with standards made by dissolving NBS organometallic compounds in the same solvent. Molecular absorption was corrected by measurement at a non-absorbing wavelength.

Trent and Slavin used the method of standard additions, adding increasing amounts of nickel cyclohexane butyrate, dissolved in 'pure' lubricating oil and diluted in *p*-xylene, to thin oil samples also diluted with xylene. An air acetylene flame with auxiliary air was used. Kerber used a 1 + 4 dilution in *p*-xylene, constructing a calibration curve from synthetic standards made from the same materials, and correcting for background absorption. A detection limit of 0.05 p.p.m. of nickel was obtained.

Moore et al.[277] found that dioxane was a better solvent for crude and fuel oils from the point of view of solvent power and burner characteristics. They determined copper and nickel using the standard addition method.

Vanadium was determined by Capachi-Delgado and Manning[82] with a nitrous oxide acetylene flame. With a 1 + 2 dilution in *p*-xylene, a detection limit of

0.05 μg/ml was obtained. Standards were made by dissolving bis-(1-phenyl-1,3-butane diono)oxovanadium IV in oil, and diluting with the same solvent.

In the analysis of crude oil, the sampling procedure itself requires some care. Before the analytical samples are measured out, the bulk sample should be thoroughly shaken, and heavy crudes warmed to 60°C in a water bath to improve mixing. A procedure recommended for the determination of nickel, sodium and vanadium at 1–200 p.p.m. in light and heavy crudes is as follows:

> Dilute samples 1 + 9 in a mixture of 10% isopropanol and 90% white spirit. Prepare standards by dissolving appropriate amounts of nickel, sodium and vanadium naphthenates in the same solvent. Only stock standards e.g. 500 p.p.m. or 1000 p.p.m. should be stored. Small quantities of working standards are prepared afresh daily, containing for example 1–10 p.p.m. of nickel, 0.5–5 p.p.m. of sodium and 5–50 p.p.m. of vanadium. For higher concentrations, further dilute the samples.
>
> The use of the nickel line 341.4 nm instead of 232.0 nm avoids the need for scatter correction. Sodium is measured at 589·0 nm and vanadium at 318·4 nm. Although metal naphthenates are economical and perfectly suitable for use as standard material, the composition from batch to batch may vary somewhat, and the contents of a new bottle should consequently be assayed by dry- or wet-ashing followed by conventional methods.
>
> White spirit (petroleum hydrocarbons, mainly decane, boiling between 150 and 200 ° C) is the preferred solvent, as it has approximately the same viscosity as water, and it is found that better reproducibility can be obtained than with *n*-heptane or iso-octane. The stability of the flame and its burning characteristics are improved by the addition of 10% of isopropanol.

Lubricating Oils and Greases. While trace elements are determined by methods similar to the above, additives in lubricating oils and blends may be determined after greater dilutions. Mostyn and Cunningham[291] used a 200× dilution in *n*-heptane to analyse blends for zinc. Standards were made by diluting either the actual additive or zinc naphthenate with the same solvent.

Ishii et al.[183] also determined zinc but preferred methyl isobutyl ketone as the solvent. Standards were made by dissolving lead nitrate in this solvent, with addition of liquid paraffin in order to match the viscosity. Kohlenberger[214] examined the effect of various solvents on the determination of calcium and zinc concluding that higher precision was obtained when higher boiling point solvents such as white spirit were used. Schallis and Kahn[349] used the nitrous oxide acetylene flame for determining tin, and Pickles and Washbrook[295] used a relatively simple equipment for determining both trace metals and barium, calcium and zinc additives, the latter all with nitrous oxide acetylene. Potassium, as ionization buffer, was added at the 1% level in the form of potassium alkyl sulphonate, and the solvent employed was white spirit.

The use of the nitrous oxide acetylene flame is strongly recommended for these three added metals. Barium gives better sensitivity, and there are no interferences from phosphorus on either barium or calcium. Even zinc, for which a cool flame is normally used, can be determined with adequate sensitivity, but more important, the calibration graph is found to be less curved, and the burner change is avoided. It is necessary to add an ionization buffer. This can be potassium as noted above, or sodium added simply as sodium naphthenate to a level of about 500 p.p.m. The general solvent, 10% isopropanol/90% white spirit, is also used for this analysis.

Greases may be treated by one of two methods—dissolution in a suitable solvent, or, if the elements to be determined are at too low a concentration level, dry-ashing followed by dissolution of the residue in hydrochloric acid. Mostyn and Cunningham[281] used the latter method even for determining the molybdenum sulphide content of lubricating greases, finding it to be both more rapid and more convenient. Ashing would have to be carried out at a carefully controlled temperature to avoid loss of elements by volatilization. In particular, if molybdenum is being determined, the temperature should not exceed 600°C. The ash was extracted with ammonia and the samples and standards were made up to contain equal concentrations of ammonium chloride.

Wear Metals in Lubricating Oils. An increase in the metal content of crank-case and circulated lubricating oils signifies a source of potential trouble in an engine. The actual metal may indicate the location of the trouble. For example, the presence of lead, tin or antimony might mean wear in a bearing, iron, chromium or nickel would be present because of piston wear, while contamination by sodium or boron could arise because of a leak in the cooling system. Periodical analysis of the lubricants enables such troubles to be located and corrected before they cause a major breakdown.

Wear metals are different from naturally occurring trace metals, in that they are probably present in the form of fine metallic particles, or even of a colloidal suspension. They may therefore be expected to give a different response from the same metals present in the form of soluble organic salts as in oils with additives or in working standards. Experience has shown, however, that the particle size is frequently so small that, with hot flames, the metal is completely atomized and hence good results can be obtained in relation to synthetic standards made up from organometallics.

Means and Ratcliffe[265] successfully determined iron, silver, copper, magnesium, chromium, tin, lead and nickel in jet engine lubricants. After a 1 + 4 dilution with *p*-xylene the oil was aspirated into an air acetylene flame. Standards were of metal naphthenates dissolved in 'unworn' oil, also diluted with *p*-xylene. Methods for preparing metal naphthenates were given. Burrows[78] in a similar method used methyl isobutyl ketone as the solvent for railway engine oils and Sprague and Slavin[372] described a rapid method for aircraft engine oils. By utilizing a composite standard solution for seven elements, iron, nickel, chromium, lead, copper, silver and magnesium and a digital concentration readout system, a precision of 5% was claimed for an analysis rate of fourteen seconds per element per sample. The method was soon made fully automatic.[361] One hundred oils can be analysed for nine elements in four hours.

One of the particular problems in analysing this type of oil is again in taking a truly representative sample. The metallic particles gradually separate on standing, and settle to the bottom of the container as a sludge. The main sample should therefore be taken from the engine quickly after shut-down, and should be re-warmed and thoroughly mixed again before the working samples are measured out. As before, a dilution of 1 + 4 with 10% isopropanol/90% white spirit mixture is the recommended procedure, and standards are made by dissolving metal naphthenate in the same solvent. It is probably unnecessary to add unused oil to the standards, for the viscosity difference is negligible, and in any case, the important feature of this analysis is that it enables concentration *trends* to be quickly recognized over specified periods of time. Provided the figures are correct relatively, an actual error of 5% or even more is usually unimportant.

Analysis of Fuels. Trace elements in both fuel oils and gasolines often require to be determined, particularly lead, copper and zinc, as they accelerate the oxidative deterioration of refined products or otherwise reduce their storage stability, and vanadium which, when present in oil destined for boiler firing, leads to corrosion and produces a fine toxic ash.

Scott and Killer[352] used methyl isobutyl ketone as the solvent for direct determination of vanadium in fuel oil, diluting the sample 1 + 9. They suggested that the dilution technique provides sufficient sensitivity for trace elements at concentrations down to about 1 p.p.m. Below this they recommended that the trace metals be separated and concentrated by wet-ashing, chromatography or extraction. Acid extraction would appear to be a useful technique, as 90% or greater recoveries of vanadium, nickel and iron were obtained when 250 ml of artificial blends were shaken with 10 ml of 2 M-hydrochloric acid.

The petroleum analysis causing the greatest problems in atomic absorption is the determination of lead, added in the form of tetraethyl lead (TEL) or tetramethyl lead (TML) as an anti-knock. Its concentration is usually between 200 p.p.m. and 1500 p.p.m. and it may be present as either TEL or TML or as a mixture of both. Standards produced by dissolving TEL in gasoline generally give a calibration graph of different slope from that produced with TML. It must therefore be known which compound is present, but if both are suspected a correct result is unlikely to be obtained.

Trent[401] attempted to minimize the difference between the TML and TEL lead absorption responses by careful choice of solvent and by introducing the sample to the spray chamber through a long capillary tube in order to decrease the uptake rate and improve the nebulization efficiency. Mostyn and Cunningham[281] confirmed that this problem persists, though, in limiting their work to materials containing only TEL, they obtained some very satisfactory results in spite of a 'memory' effect in which TEL standards take more than 90 seconds to reach a maximum reading. The problem was overcome by using a carefully timed operating sequence of fifteen seconds aspiration of the gasoline sample or standard, followed by $1\frac{1}{2}$ minute interval wash of pure iso-octane. A coefficient of variation of 1.6% was obtained at a concentration level of 3.5 ml of TEL per imperial gallon of gasoline. 2 ml of sample are diluted to 100 ml with iso-octane. Standards are prepared from TEL and iso-octane.

Total lead in gasoline may be determined after a simple and rapid extraction with iodine monochloride.[188] Lead is removed from both alkyls, the final solution is aqueous, and comparison standards consist simply of dilute solutions of lead nitrate. The method is as follows:

> Pipette 5 ml of gasoline into a 100 ml separating funnel containing 10 ml of 1.0 M-iodine monochloride. Add 25 ml of iso-octane, stopper the flask and shake for 3 minutes. Transfer the lower, aqueous, layer to a 100 ml calibrated flask. Wash the remaining organic layer with 10 ml of water by shaking for a further one minute and add the washings to the flask. After diluting to volume with water, aspirate into an air acetylene flame. Compare the absorbance with that given by aqueous standards

of lead nitrate containing 5–50 p.p.m. of lead. Good sensitivity is obtained by using the lead resonance line at 217.0 nm.

Iodine monochloride is simply prepared for this determination:

Add 112 ml of 25% potassium iodide solution to 112 ml of hydrochloric acid sp.gr. 1.18, in a 500 ml beaker, and cool to room temperature. To the cooled solution add slowly 18.75 g of potassium iodate crystals and mix until a clear orange coloured solution is obtained. Dilute to 250 ml with water.

It was reported by Kashiki et al.[198] that lead, present as TEL, TML, or as mixed alkyl leads, can be determined in a premixed air acetylene flame against a single alkyl lead standard if 3 mg of iodine are added to the 1 ml petrol sample. This is then diluted to 50 ml with methyl isobutyl ketone. It was stated that there are no interferences and the method is suitable for determining from 0.16 to 0.8 g of lead per litre of gasoline.

Paints, oils and colours

Metallic elements occur as principal constituents in inorganic pigments and in additives to drying oils, but may be present in smaller amounts in pesticides and antifouling compounds. Trace elements may be undesirable because of their toxic properties or because they affect the colour or stability of the product.

Regulations are laid down in some countries on the toxic metal content of paint for toys and domestic appliances. In the United Kingdom for example, the 'soluble' lead content of a dry coating of paint on a toy must not exceed 5000 p.p.m. and arsenic, antimony, barium, chromium and cadmium must each not exceed 250 p.p.m. The soluble metal content is that which can be extracted in a 0.25% solution of hydrogen chloride under specified conditions, which approximates to the dissolving power of human gastric juices.

There have been very few references to the analysis of paint and related materials by atomic absorption since an early paper by McPherson.[264] Both paints and plastics were dissolved in an appropriate solvent, and sprayed directly into an atomic absorption spectrometer.

The problems encountered[307] are not unlike those in the petroleum industry, except that the vehicle may consist of natural or synthetic polymerizable fluids and the pigment is a suspended solid of high concentration. Many of the solvents used in the analysis of petroleum products viz. methyl isobutyl ketone, *n*-heptane, iso-octane or white spirit, can equally well be used to dissolve paint vehicles. White spirit (which is also used as a thinner in the paint industry) is preferred for reasons already given in the section on petroleum.

Direct Methods. Oils or thinners to be analysed for trace elements may be diluted with white spirit by a factor of ten, and compared with standards of metal naphthenates also dissolved in white spirit.

Drying oils themselves often contain metal naphthenates, especially those of cobalt, lead, manganese, zinc or zirconium, and these metals are all determinable by a similar method, though a greater dilution factor would be required.

Pretreatment Methods. To conform to the official specification,[373] 'soluble' lead, arsenic, etc., are determined in the aqueous phase after a dried and weighed quantity of the sample is shaken for one hour at 16°C with 1000 times its weight of hydrochloric acid solution containing 0.25% by weight of hydrogen chloride. Standards would be made up to contain the same concentration of hydrochloric acid.

Pigments may be separated to enable either the pigment itself or the vehicle to be analysed, by dissolution of the vehicle in an organic solvent followed by centrifugation. Inorganic pigments are then dissolved in a suitable acid mixture, or if necessary first fused with an alkali salt, and then leached out with acid. The vehicle is analysed by a direct method as described above, unless the metal concentrations are too low to be determined direct.

A useful means for bringing organic materials into aqueous solution in readiness for atomic absorption analysis is the oxygen flask method[256] which has been applied to the analysis of commercial plastics. More usually, however, the organic constituents of paints can be analysed for very low concentrations of trace metals by a preliminary wet oxidation step. Powder paints, too, with organometallic constituents, and antifouling paints, which contain compounds of copper and mercury are best treated by wet oxidation. However, phenyl mercury compounds are volatile, and if these are present, the procedure must be carried out carefully to avoid loss of mercury. Suitable wet oxidation methods for organic materials with or without mercury are given by the Society for Analytical Chemistry.[191]

With the analyte in aqueous solution it is sometimes desirable to separate the trace elements, either from an excess of alkali fusion salt or from a major pigment element. One of the separation techniques described in an earlier section may then be used. Probably the ammonium pyrrolidine dithiocarbamate–methyl isobutyl ketone system is the most useful.

Major Constituents. Separated inorganic pigments can in principle be checked for major or impurity elements. Many pigments dissolve readily in one or other of the mineral acids. Titanium dioxide may cause some difficulties, however, and is dissolved in sulphuric acid after fusion with potassium bisulphate and glucose. Antimony has been determined in titanium dioxide pigments by this method[269] after subsequent extraction into methyl isobutyl ketone.

The main pigment elements include aluminium, antimony, arsenic, barium, calcium, cadmium, copper, chromium, mercury, iron, lead, magnesium, manganese, sodium, potassium, silicon, titanium, zinc and zirconium. The sensitivity of many of these may need to be reduced by burner rotation or measurement of an alternative resonance line. Operating conditions may be inferred from the details given in Appendix 1.

Textiles, fibres and plastics

Both natural and synthetic fibres have been analysed for trace elements, but synthetic fibres may also contain residues of catalysts, stabilizing agents etc.

Aluminium was determined in wool by Hartley and Inglis[165] after dissolving the sample in constant boiling mixture hydrochloric acid. Although hydrochloric acid tends to depress the response of aluminium in a nitrous oxide acetylene flame, the presence of wool protein aminoacids enhance it. Standards were therefore prepared to contain the appropriate acid concentration and hydrolysed wool protein. Hair and hide could also be analysed by the same method.

Copper was determined quickly and reliably in textile materials after simple acid extraction by Simonian,[358] but synthetic fibres were first ashed by Yanagisawa et al.[457] who extracted germanium from the residue, dissolved in hydrochloric acid, with methyl isobutyl ketone. Other ions were extracted simultaneously (iron, antimony, arsenic and gold) and presumably these could be determined by the same technique.

Trace metals in synthetic fibres and polymers remaining from previous processing may affect, either adversely or favourably, their stability and other characteristics. Olivier[285] evolved methods for the examination of four types of material: soluble polymers; insoluble polymers; wool, cotton and cellulose; polymers containing volatile metallic compounds. Polymers (0.5 g sample) are dissolved, where possible, in an organic solvent. Appropriate solvents are given in Table 13.

TABLE 13

Polymer	Solvent
Polystyrene	Methyl isobutyl ketone
Cellulose acetate	Methyl isobutyl ketone
Cellulose acetate butyrate	Methyl isobutyl ketone
Polyacrylonitrile	Dimethyl formamide
Polycarbonates	Dimethyl acetamide
Polyvinyl chloride	Dimethyl acetamide
PVC/PV acetate	Cyclohexanone
Polyamides (nylon)	Formic acid
Polyethers	Methanol
Wool	5% Sodium hydroxide
Cotton and Cellulose	72% Sulphuric acid

After warming if necessary the solution is made up to 25 ml with the solvent and aspirated direct.

It is important to ensure that, if organic solvents are to be aspirated directly, they will not dissolve or otherwise damage parts of the sample-handling section of the atomic absorption spectrometer that are themselves constructed from plastic or polymerized materials.

Polymers which do not dissolve in an organic solvent must be wet-ashed. The sample is heated with 2–3 ml of concentrated sulphuric acid, and hydrogen peroxide added dropwise until the organic matter has been destroyed. The solution is then made up to the required volume with water.

Wool, cotton and cellulose fibres are hydrolysed, again taking 0.5 g samples. Wool is treated with 15 ml of 5% sodium hydroxide solution, 'dissolution' being complete in about 30 minutes. If the sodium is likely to interfere, hydrochloric acid may be used instead, but this takes longer. Cotton and cellulose are shaken for 30 minutes with 6–7 ml of 72% sulphuric acid. Finally, in all cases the solution is made up to 25 ml with water.

Volatile metal compounds are determined after the polymer has been dissolved in any possible solvent. Where the solvent is unsuitable for atomic absorption analysis (organic halides, cresols—see page 84—concentrated acids, etc.) a double precipitation into aqueous medium is carried out. The organic solution is dropped, slowly with vigorous stirring, into water and after the second such precipitation, the metals are transferred quantitatively to the aqueous medium. For lead, the aqueous medium should be dilute ammonium acetate, for silver dilute ammonia, and for antimony dilute hydrochloric acid. A further concentration of most metals can then be made with the ammonium pyrrolidine dithiocarbamate/methyl isobutyl ketone system, or with cupferron for aluminium extraction.

Plastic materials may be treated either by the oxygen flask method of Matsuo[256] or ashed.

The following wet-ashing procedure has been found to be suitable for determining calcium and magnesium in polyvinyl chloride:

> To 1 g of sample add 20 ml of perchloric acid. Heat gently and then evaporate to fumes. Add 20 ml of 5% lanthanum chloride solution, filter, and make up to 100 ml. Compare with standards containing 0–10 p.p.m. of calcium and magnesium and 1% lanthanum chloride, aspirating into an air acetylene flame.
>
> If any residue remains on the filter paper, ash the paper and fuse with 5 g of potassium hydroxide. Extract the fusion with 5 ml of hydrochloric acid, add 20 ml of 5% lanthanum chloride solution, and make up to 100 ml.
>
> If calcium or magnesium are present, compare with standards containing 0–10 p.p.m. of calcium and magnesium, 1% of lanthanum chloride and 5% of potassium chloride.
>
> Lanthanum chloride is added when an air acetylene flame is used, as there is a possibility that titania or alumina fillers may pass into solution to a sufficient extent to cause some depression of the calcium absorption response.

Organometallic compounds

The analysis of organometallic compounds may include the determination of the metallic components by atomic absorption. For independent analyses the compounds are wet-ashed, and the resulting aqueous solutions are compared with synthetic aqueous standards.

A typical procedure suitable for the determination of arsenic in triphenyl arsine and phenyl arsonic acid, and for the determination of barium, calcium, cadmium, cobalt, manganese, lead and zinc in organometallics is:

> Dissolve 0.5 g samples in 20 ml of 1 + 1 nitric perchloric acid mixture. Digest gently for four hours and dilute to 100 ml with water. Make a further dilution (e.g. 1 + 19 for the arsenic compounds mentioned) and compare with standards containing 10–200 p.p.m. of arsenic or barium or 1–20 p.p.m. of other elements in perchloric acid of the same concentration as the prepared sample solution.

This procedure is not suitable for the determination of tin as metastannic acid is precipitated. To determine tin in, for example, dipropyltin dichloride or hexabutylditin:

> Warm carefully 0.1 g samples in 15 ml of 1 + 1 + 1 nitric–hydrochloric–phosphoric acid mixture under a reflux condenser. When dissolution is complete, dilute to 100 ml with water. Compare with standards containing 10–200 p.p.m. of tin in an acid mixture of the same final concentration.

Analyses of materials containing known stable organometallic compounds may be done by a direct procedure using standards carefully made up from the compound itself. The method can be used, for example, in the control of the composition of silicone fluid mixtures and blends. Samples may be diluted in ethanol or a suitable ethanol water mixture, while standards are made up by dissolving the silicone fluid in the same solvent.[125]

Pharmaceuticals and toiletries

Both major and unwanted trace elements have to be controlled in pharmaceutical products, the individual components or raw materials.

Direct analysis is possible if the material is soluble in aqueous or non-aqueous media, and standards are made up in the same solvent and with the same major constituents present.

Concentration of trace elements, if necessary, is achieved by the methods already described, solvent extraction or wet oxidation, the latter always being relied upon if an entirely independent method is needed. Lead, arsenic, mercury, copper, iron and zinc may have strictly enforced upper concentration limits, and these elements may be introduced in the raw materials or through contamination from the equipment used in the preparative processes. Lead, arsenic and mercury do not have good sensitivities and special conditions may have to be used for one or all of these elements (see Appendix 1). Leaton[232] described procedures for the determination of metals in drugs, dealing with liquids, suspensions, tablets, powders and ointments. Prepared samples were diluted with either 2% nitric acid or 1% lanthanum chloride solution.

The determination of zinc in crystalline zinc insulin and its preparations was described by Spielholtz and Toralballa.[368] 5 mg of sample is suspended in 10 ml of water and dissolved by the addition of one drop of hydrochloric acid. This solution is then made up to 50 ml. Alternatively 1 ml of protamine insulin solution is diluted to 50 ml, after acidifying with 1 drop of hydrochloric acid if necessary. Standards in the range 0.2–3.0 p.p.m. of zinc were prepared simply by diluting a stock zinc solution with water. The addition of albumin had no effect on the zinc absorbance. Zinc insulin is used in therapeutic medicine, but zinc and other heavy metals, which would be determinable by a similar method are used in x-ray studies of the structure of insulin.

Vitamin B_{12} (cyanocobalamin) in pharmaceutical dosage forms was determined indirectly by its cobalt content by Berge et al.[54] About 20 tablets were dissolved in hot water, the solution filtered, and cobalt determined in the filtrate. Comparison was made with a standard solution of cyanocobalamin.

The determination of aluminium, bismuth, calcium and magnesium in antacid preparations presents an almost classical example of stable compound interference. Aluminium itself must be determined in a nitrous oxide acetylene flame, and bismuth in a lean air acetylene flame. The interference of aluminium on magnesium and calcium in the air acetylene flame is completely overcome by addition of lanthanum. The procedure is as follows:

Dissolve 0.5000 g of powdered sample in 10 ml of water, 5 ml of hydrochloric acid and 5 ml of nitric acid. Digest on a hotplate until dissolution is complete. Transfer to a 100 ml flask, washing with 20 ml of M-hydrochloric acid and dilute to the mark. Dilute this solution further as indicated in Table 14 incorporating lanthanum chloride solution in the final analysis solution to give a concentration of 1% lanthanum where appropriate in both samples and standards.

The further dilutions of the first solution are necessary to accommodate the different sensitivities of these four elements. Silicon is another element which is present in some preparations in major concentrations, either as 'tri-silicate' or

TABLE 14

	Further dilution of sample solution	Standard range	Lanthanum conc. in final soln.	Equiv. % in sample
Aluminium	5 ×	0–100 mg/l	–	0–10% Al
Bismuth	—	0– 80 mg/l	–	0– 1.6% Bi
Calcium	50 ×	0– 25 mg/l	1%	0–25% Ca
Magnesium	500 ×	0– 2 mg/l	1%	0–20% Mg

even as silicone oil. In the latter case, unfortunately, homogenization with ethanol-based diluents (as mentioned for silicone oil blends, p. 153) does not appear to be successful, and therefore preliminary cold wet oxidation (avoiding loss of silicone) must be applied before taking up in a hydrofluoric acid-based solvent. Such a solvent is also used, in methods similar to those described for siliceous materials, for trisilicate materials.

THE ANALYSIS OF BIOLOGICAL SAMPLES

In this section the 'biological' label is intended to cover all materials—organisms, plants, animal products—that have been formed as a result of, or are necessary to support, various forms of life. Medical and pathological applications of atomic absorption are the subject of the next section. The recent book by Christian and Feldman[8] is an excellent guide to the literature relevant to the total biological field up to about mid-1969.

The major problem in the analysis of biological samples, whatever their origin, is the variety of forms in which the organic matrix may exist. This may be overcome in a widely applicable method in which all biological samples are either dry- or wet-oxidized, and the residue taken up in some mineral acid. After addition of a spectroscopic buffer (e.g. lanthanum chloride if an air acetylene flame is to be used, potassium or lithium chloride if nitrous oxide acetylene) the solution would be made up to volume and compared with standards containing the same anions and the same buffer. This is the basis of very many methods that have been published during the past ten years, and where an unusual analysis has to be carried out, without preliminary research, the principle can be relied upon to give accurate results.

More direct methods, e.g., dissolution of the sample if possible, acid extraction of the wanted metals, are less time-consuming, but have to be proved to give correct results for each different matrix, and may not, without further operations, give adequate sensitivity for trace elements.

We consider biological samples according to the type of matrix, for in accordance with principles already established, it is invariably possible to determine more than one element, if desired, in a given prepared sample.

Liquid samples—waters, juices, beverages

Table and mineral waters would be analysed by methods already given for natural waters, except that it may be necessary to remove dissolved gases such as carbon dioxide as these may interfere with the regular flow of sample through the nebulizer capillary.

Fruit Juices. These are analysed for trace elements absorbed from the environment or for elements such as iron and tin which derive from the cans in which juices are stored and distributed. The problem of iron and tin was studied by Price and Roos[311] who showed that, for orange and pineapple juices, correct results are obtained by acidifying the juices and separating suspended matter by centrifugation. The method was extended[331] to other metals in juices: calcium, magnesium and potassium—which are present in relatively major concentrations—and manganese, zinc and sodium. A 5× dilution of the sample is necessary to overcome the viscosity effects caused by the presence of sugar, while hydrochloric acid both extracts metals from suspended solid material and also standardizes the acidity of the prepared solution. The procedure is as follows:

> To 20 ml of the juice in a 100 ml volumetric flask add 10 ml of hydrochloric acid, then make up to the mark with water. Shake well, transfer sufficient quantity to a dry centrifuge tube and centrifuge. This solution is aspirated for iron and tin, but further dilutions as shown in Table 15 may be needed for the other elements. Make dilutions so that the final solution always contains 5% of hydrochloric acid, and the solution for calcium contains 0.5% of lanthanum as lanthanum chloride. Prepare standards containing the same acid and lanthanum concentrations as the samples. For tin use the nitrous oxide acetylene flame, and for all the other elements air acetylene.

The method was tested both by subjecting a number of samples also to wet oxidation, comparing with 'pure' aqueous standards, and by determining recoveries of known amounts of each element added to the samples. Interferences were detected in the air acetylene flame between iron and the organic constituents of the juices, citric acid and sugar. In test experiments this was completely overcome by the addition of 60 p.p.m. of phosphoric acid, less than is naturally present in the juices. There is thus no need to make special additions to the juices or standards to overcome this effect. None of the other elements is affected in this way, and there is no interference with tin in the nitrous oxide acetylene flame. Lanthanum is added to prevent depression of the calcium absorbance by phosphate.

Typical tin and iron contents of fresh, bottled and canned juice are given in Table 16 and indicate that the range of standards for tin in the canned product may have to be extended. The simplicity of the direct atomic absorption method

TABLE 15. Preparation of fruit juices

	Approx. level p.p.m.	Dilution	Standard range p.p.m.
Tin	*	none	0–50
Iron	*	none	0–10
Manganese	5	5 ×	0–10
Zinc	5	5 ×	0–10
Calcium	100	10 ×	0–25
Sodium	20	10 ×	0– 2.5
Magnesium	200	200 ×	0–10
Potassium	1000	200 ×	0–20

* See Table 16.

TABLE 16. Typical tin and iron contents of fruit juices p.p.m. (Price and Roos[311])

	Tin	Iron
Fresh orange juice	7.5	0.5
Bottled orange juice 1	25	2.0
2	45	2.5
Canned orange juice 1	60	2.5
2	115	0.5
3	120	2.5
Bottled pineapple	45	15
Canned pineapple	130	17.5

makes it ideal for quality checks by manufacturers and health authorities, but if other trace elements are to be determined (notably lead, which is less than 0.5 p.p.m. in the undiluted juice) a concentration and/or extraction step would have to be carried out.

Wines and Beers. The significance of traces of some metals in wines and beers is well recognized. The stability of beers, and also the quality of the foam, is associated with the presence of some metals, while in the U.K. the concentration of lead, copper and zinc is the subject of legislation. Zinc-deficient worts may have to be supplemented with zinc ions in order to stimulate fermentation.

Probably the first application of atomic absorption to the analysis of wines was made by Zeeman and Butler[459] who dehydrated the samples with sulphuric acid and then ashed at 500°C before taking up again in acid. Frey et al.[144] determined a number of trace metals in beer, wort and yeast either direct or after extraction, and subsequently[145] investigated the effect of several of these metals on the fermentation processes. Strunk and Andreasen[377] aspirated alcoholic beverages directly in order to determine copper, while Meredith et al.[270] in the same laboratory found it necessary to add ethanol to the standards, bringing them to either 86° or 100° proof in order to determine iron in beverages of similar proof strength. Ethanol was added to wine samples, normally of 40° proof, in order to conform with the composition of the standards. Beers were not included in their investigations.

The determination of copper, iron, zinc, lead and magnesium in both wine and beer for routine quality control was described by Weiner and Taylor.[421] These elements are mostly determined by direct aspiration, though the lead content is normally so low that extraction is necessary, and iron also may have to be extracted from beer and some table wines.

The following methods, based on Weiner and Taylor, will be found to be satisfactory for beer and table wines:

Direct method: decarbonate beer by pouring rapidly from one beaker to another a number of times and allow to settle so that the foam collapses back into the liquid. Aspirate beer and wine samples without further treatment unless dilution is seen to be necessary. Prepare standards by adding up to 1 p.p.m. of copper, 2 p.p.m. of zinc and 2 p.p.m. of magnesium to decarbonated beer or wine of known low trace metal content. Aqueous standards are prepared for wines; these should contain 10% of ethanol.

Extraction method: transfer 50 ml of wine or decarbonated beer to a 250 ml beaker, add 5 ml of glacial acetic acid and boil for two minutes. Cool and transfer quantitatively to a 100 ml centrifuge tube. Add 2 ml of 1% aqueous ammonium pyrrolidine dithiocarbamate solution and, by pipette, 10 ml of methyl isobutyl ketone. Shake for about 5 minutes and centrifuge at 3000 rev./min for 10 minutes. Aspirate the upper organic layer direct from the tube. Prepare standards by pipetting 0–5 ml of standard iron (50 mg/l) and standard lead (10 mg/l) solutions into 100 ml calibrated flasks, and make up to the mark with water. Pipette 50 ml of each solution into 100 ml centrifuge tubes, add 5 ml of glacial acetic acid, 2 ml of 1% ammonium pyrrolidine dithiocarbamate solution and, by pipette, 10 ml of methyl isobutyl ketone. Continue as for the samples. These standards are equivalent to 0–2.5 mg/l iron and 0–0.5 mg/l of lead in the original samples. If only iron is to be determined the addition of acetic acid and the boiling can be omitted.

The response of both copper and iron may be enhanced in the presence of ethanol. This is offset by use of 'pure' beer or wines where possible in preparation of standards for the direct methods and, of course, is eliminated in the extraction method. This finding is consistent with those of Meredith et al.[270] and beverages of higher ethanol content should be compared with standards carefully matched with respect to ethanol when the direct method can be employed.

Boiling with acetic acid breaks down possible lead complexes in beer, so that small amounts can be completely recovered. The air acetylene flame is used for all of these elements, though a small improvement in sensitivity may be achieved for copper and zinc, if required, with air propane.

There is no doubt that other extractable elements can also be determined in the organic phase in the second procedure.

Food and feedingstuffs

Convention has it that human beings eat food while animals feed. Both food and feedingstuffs consist mostly of an organic matrix containing traces of various metals. In feedingstuffs these are often added in the form of mineral mixes while in foods they can be present as desirable minerals or undesirable toxic elements. Either way their concentration has to be known and controlled, but in atomic absorption the analytical problems are identical. The metals must be separated from the solid matrix and this may be achieved either by removing the matrix by wet or dry oxidation or by extracting the elements into mineral acid.

A collaborative study on the determination of zinc in foods was reported by Rogers.[328] Both wet- and dry-ashing procedures were examined, and 100% recoveries were obtained in each case. Dalton and Melanoski[112] determined copper and lead in meat and meat products. Samples were dry-ashed at 500°C and the residue dissolved in nitric acid. Copper and lead were determined in the range 1–5 p.p.m. and 1–10 p.p.m. respectively. The lithium contents of some consumables—salt, lettuce, potatoes etc. were determined by Hullin et al.[178]

There are one or two non-typical examples of non-ashing methods. Food grade phosphates were analysed for lead after dissolution in hydrochloric acid and extraction with diethyl-ammonium dithiocarbamate into xylene.[192] Zinc was determined in sugar products and molasses solutions after further dilution with citric acid solution by Mee and Hilton.[266] This procedure gave the same results as dissolution after ashing. Hoover et al.[175] determined lead in plant and animal products by digestion with nitric, sulphuric and perchloric acids followed by co-precipitation of the lead on strontium sulphate. After conversion to lead carbonate by agitation with ammonium carbonate, the precipitate is dissolved in nitric acid and the lead determined in an air acetylene flame.

For routine analyses of both foods and feedingstuffs the dry-ashing method has two particular advantages: it requires less operator time than wet-ashing and it leaves a residue containing the minimum amount of unwanted ions. Such a residue is readily dissolved in hydrochloric acid. Dry ashing procedures must be carefully carried out to guard against the possibility of loss of wanted elements by vaporization. An ashing temperature of not more than 500°C will ensure that most elements, with the exception of mercury, arsenic and some other metalloids, remain in the residue. For the exceptions mentioned, wet oxidation is essential, and that for mercury in the cold.

Dry-ash Procedure. A typical routine dry-ashing procedure was described by Roach et al.[325] and by Williams.[434] The method is used for the analysis of raw materials and for the control of formulated feedingstuffs for added copper, zinc and magnesium. Recovery experiments indicate that none of these elements is lost during the procedure described, and the accuracy of the method was established by comparison with standard colorimetric methods. Coefficients of variation of 1–2% were obtained for copper and magnesium and 3% for zinc, all in the p.p.m. range.

The procedure is readily extended to other elements, e.g. calcium and manganese. Recommended operating details are as follows:

Grind the sample if necessary so that it passes a B.S.S. No. 16 sieve. Weigh 2.0 g of sample (or other convenient quantity) into a silica evaporating basin, place on a hot-plate and allow to 'smoke' until completely charred. Transfer the basin to a muffle furnace at a temperature of 470° C and ash.

When ashing is complete, cool and extract with the minimum amount of hydrochloric acid and evaporate to dryness. Extract again with 10 ml of 25% hydrochloric acid, boil and filter into a 100 ml calibrated flask. Wash the filter through with warm 1% hydrochloric acid solution, make the contents of the flask up to 100 ml with water and mix.

This solution is used for the determination of copper, magnesium, manganese and zinc. To determine calcium, dilute the solution by a further factor F to give a calcium concentration in the range 1–10 mg/1, adding 2 ml of 5% lanthanum solution per 100 ml of final solution. An air acetylene flame is used for all these elements. Multi-element standards are prepared to contain 0–10 mg/1 of copper (equivalent to 0–0.05% in an original 2 g sample) 0–2 mg/l of magnesium (0–0.01%) and 0–5 mg/l of both manganese and zinc (0–0.025 %).

Standards for calcium should contain 0–10 mg/l of calcium and 0.1 % lanthanum as lanthanum chloride. The calcium content of the original sample would be $VF/100W$ % where V is the reading in mg/l from the calibration graph, F is the further dilution factor, and W is the weight in g of the original sample.

This method has been applied successfully to all types of animal feedingstuffs, oilseed cakes, and to milk, bonemeal and other products of animal origin. There is no obvious reason why it should not be applied to any material with a largely organic matrix and in this sense it may justifiably be described as universal.

Acid Extraction. For the vitamin–mineral mixes actually added in the preparation of feedingstuffs, a simple acid extraction technique may be used. The following may be used in order to determine cobalt, copper, iron, magnesium, manganese and zinc in this type of sample:

Extract 1 g of sample with 20 ml of hydrochloric acid at 65° C for 30–60 min, adding further acid if necessary. Then add a little water, boil and filter into a 50 ml calibrated flask and make up to volume. Compare with standards made up in 5% hydrochloric acid, using an air acetylene flame.

Acid extraction may also be used to determine trace elements in water-insoluble matrices. A good example of this type of sample is butter or edible oil. The latter are often produced in a process which involves a nickel catalyst. Both the nickel and copper contents usually have to be known. The simple acid extraction procedure may well result in an emulsion which it is difficult or impossible to break but carbon tetrachloride dissolves the organic phase, preventing formation of the emulsion and forming a lower layer. The aqueous layer containing the extracted elements remains on the top, ready for direct aspiration. Details (Price et al.[312]) are as follows:

Dissolve 10 ml or 10 g of sample in carbon tetrachloride in a 50 ml centrifuge tube and add 10 ml of 10% nitric acid. Stopper and shake continuously for 10 minutes. Centrifuge for 2 minutes at 2500–3000 rev./min. Aspirate the upper layer, and compare with standards containing 0–50 p.p.m. of nickel and copper in 10 % nitric acid.

Determination of Mercury in Foodstuffs etc. Fractions of a part per million of mercury in foodstuffs are determined, after cold oxidation of the sample, using the cold vapour (flameless) method. Sample preparation is the same as that given for tissue (page 174) and the measurement is described under Mercury (Appendix 1).

Plant materials

As early as 1958, David[118] determined zinc, iron, copper and magnesium in plant material. Although his equipment was less sophisticated, his preparation procedure could hardly be improved today. One or two grams of the dried plant material was digested with 4 ml of sulphuric–perchloric acid 1 + 7, and nitric acid was added in small amounts until destruction of the organic matter was complete. The mixture was then fumed until all nitric and perchloric acid was removed, and the remaining 0.5 ml of sulphuric acid solution was made up to 10 ml with water. Many further procedures originated in this author's laboratory including that for calcium[117] and strontium[114] which attains good sensitivity by removal of phosphate by ion exchange.[113]

Dry-ashing and sulphuric acid digests were employed by Bradfield and Spencer[71] to determine magnesium, zinc and copper when investigating leaf analysis as a guide to the nutrition of fruit crops. The supply of calcium, manganese, strontium and zinc to plant roots growing in soil was investigated by Halstead et al.[160] in chamber experiments by determining these elements both on plant digestate and in soil extracts and the effect of magnesium deficiency in tomatoes was examined by dry-ashing different parts of the plants and taking up in acid before analysis.[418]

Typical acid digestion and dry-ashing procedures for plant materials were given by Allen and Parkinson.[34]

> *Acid digestion*: to 200 mg of finely ground plant material in a 100 ml Kjeldahl flask, add 0.5 ml of sulphuric acid, 1.0 ml of perchloric acid (60%) and 5 ml of nitric acid. Allow the digestion to proceed slowly at first and then increase heat until sulphuric acid refluxes down the side of the flask. When destruction of the organic matter is complete, cool, transfer to a 50 ml calibrated flask and make up to volume with water.
>
> *Dry-ashing*: place 200 mg of finely ground sample, contained in a silica or platinum dish in a cool muffle furnace. Allow the temperature to rise to 450° C, and ash at this temperature for 3 hours. Cool, dissolve the residue in 5 ml of 5M-hydrochloric acid. Add a few drops of nitric acid and evaporate to dryness on a water bath. Redissolve in 5 ml of 5M-hydrochloric acid, warm, filter into a 50 ml calibrated flask and make up to volume with water.

It has also been shown by Premi and Cornfield[304] that trace amounts of copper, zinc, iron, manganese and chromium can be extracted quantitatively from plant material and organic residue simply by boiling with hydrochloric acid. Complete destruction of the organic matter is therefore not important. When prepared in the way described here, the analysis solution is ready in a much shorter time than by either dry- or wet-ashing. If the method is used in routine analysis for other elements or for different types of material it would be advisable to check initially that quantitative extraction is obtained.

> *Acid extraction*: weigh 0.5 g of dried and finely ground sample into a 250 ml beaker and add 25 ml of 6M-hydrochloric acid. Boil gently for 15 minutes, allowing the volume to decrease to about 5 ml, but not to become dry. Add 5 ml of hot water, boil, filter quantitatively into a 50 ml calibrated flask and make up to volume with water. Prepare standards for each element in 0.05M-hydrochloric acid.

Grass, straw, sewage sludge and farmyard manure were all successfully analysed by this method. It is interesting to note that potassium, sodium, calcium, magnesium and manganese are all extracted quantitatively in the cold when dried, ground samples are shaken in M-hydrochloric acid for 24 hours,[159] but that under these conditions, only 60–70% of copper, zinc, iron and chromium are extracted.

Boron may be extracted from aqueous solution with 2-ethyl-1,3-hexanediol into methyl isobutyl ketone or into chloroform. This principle forms the basis for determining water soluble boron[267] and acid-soluble or total boron[420]. The nitrous oxide acetylene flame is required, but there are no significant interferences.

There are very many papers in the literature in which atomic absorption methods are recorded as having been used as the means of carrying out a particular investigation in the fields of ecology, agronomy etc. Many can be found in the *Journal of the Association of Official Analytical Chemists* and others can be traced through the classified abstracts.

Soils

Extractable Elements. The soil chemist may require to determine either extractable nutrients, the total trace element content, or even the 'ultimate soil nutrient potential'.

Many buffer solutions are used in agriculture to release exchangeable or mobile ions and these have been described by Black.[59] In general, the buffer and extracted elements may be nebulized in the atomic absorption spectrometer, with dilution as necessary. There are many examples of the direct application of atomic absorption to the buffer extracts.

Exchangeable sodium, potassium, calcium and magnesium were determined by David[116] after extraction with M-ammonium chloride solution. 0.5 M-acetic acid was used by McPhee and Ball[263] to extract calcium and magnesium, strontium being added to the extract to overcome interference by phosphate. Acetic acid extracts were used by Ure and Mitchell[405] to determine cobalt, which was present in the range 1–50 µg. This corresponds to a detection limit of 0.05 p.p.m. if an original soil sample of 20 g is shaken with 400 ml of 0.5 M-acetic acid, evaporated twice with nitric acid and made up to 10 ml in ethanolic hydrochloric acid. Tamm's reagent (0.1 M-oxalic acid + 0.1 M-ammonium oxalate) was used by Laflamme[220] to determine free silicon in soils, and by Blakemore[60] to determine extractable iron and aluminium. In this method, 10 g of soil are heated for half an hour with 200 ml of the reagent, and the resulting solution used for analysis. The extract may have to be acidified to prevent precipitation of oxalates.

Modified Morgan's solution (0.625 M-ammonia + 1.25 M-acetic acid, pH 4.8) may be used to determine magnesium values, nebulizing after dilution and adding strontium chloride solution if aluminium or phosphorus are shown to lower the magnesium absorbance. Lead and nickel were leached from soils with M-ammonium acetate in soil pollution studies.[161,262] The resulting solution was equilibrated with 0.1 M-calcium chloride solution, centrifuged and filtered for analysis. Nadirshaw and Cornfield[283] examined four extractants for assessing particular functions or potential availability of manganese in soil: M-ammonium acetate (pH 7.0) for exchangeable and water-soluble manganese II; the same, containing 0.2% of hydroquinone to extract in addition 'active' manganese and manganese III; Morgan's reagent (0.5 M-acetic acid, 0.75 M-sodium acetate, pH 4.8); and 0.5 M-acetic acid, intended as a replacement for Morgan's solution which tended to clog a single slot burner. Results were obtained by comparison with standards made up in each extractant, and agreed well with those obtained by colorimetric procedures.

A typical procedure for ammonium acetate extraction is as follows:[34]

> Extract 10 g of air-dried soil in 250 ml of M-ammonium acetate solution on an end-over-end shaker for 1 hour. Compare with standards made up in solution of the same composition. The use of a multislot burner for the air acetylene flame is recommended.

Total Content. The total mineral content of soils is not often required—particularly by the ecologist. When it is, the preparation of the initial solution requires fusion and/or hydrofluoric acid digestion procedures, depending on the elements to be determined. After the organic material has been destroyed, the methods employed would be identical with those described for the analysis of rocks and minerals, though for calcareous soils with low clay content, digestion with hydrochloric acid is generally adequate to bring nutrient elements into solution.

Biological processes

Atomic absorption is an ideal technique for the control of nutrient trace metals in processes where metal-dependent enzymes or bacteria are active. Frequent checks of a number of such metals may well be required with a good degree of accuracy so that their concentrations may be kept manually or automatically within certain closely defined limits.

A further example of this sort of application is the check made by Weiner and Taylor[421] on the magnesium content of mashing liquors and worts in beer making. The changes occurring in the magnesium content of the wort during the fermentation process was also investigated, and sufficient magnesium to ensure continued growth and metabolism of the yeast was found to be contributed by the malt itself.

PATHOLOGICAL AND MEDICAL

While pathological and clinical samples are essentially no different from other organic materials, considerable effort has been made to streamline the preparation methods. This is possible because types of sample are limited mainly to urine, serum or whole blood, cerebro-spinal fluid and tissue, and desirable because the technicians in the average laboratory have too little time for lengthy and intricate methods. Indeed, because of the comparative simplicity and undoubted reliability of atomic absorption in this field many analyses are possible which with other methods would have been time-consuming and uneconomic.

Similar methods to those described here may be applied in the field of veterinary pathology.

The elements routinely determined in serum and urine are the electrolyte metals: calcium, magnesium, sodium and potassium, and iron, copper and zinc. Lead and mercury may have to be determined in special cases such as poisoning or the screening of workers exposed to the dangers of poisoning. The normal level in serum and urine of these and other elements of interest are summarized in Table 17. All of these elements can be determined in the air acetylene flame, and even though it may not always be the most sensitive. in order to avoid the

TABLE 17. Normal levels of some elements in body fluids etc.

	Blood		Urine	Tissue
	Serum	Other fraction*		
Calcium	4.6–5.2 mequiv./l 9.3–10.3 mg/100 ml	—	5–22 mequiv./day 100–450 mg/day	—
Magnesium	1.8 mequiv./l 2.15 mg/100 ml	—	8–25 mequiv./day 100–300 mg/day	bone ash 0.6 %†
Potassium	3.1–5.5 mequiv./l† 12–22 mg/100 ml†	P 16.07 mg/100 ml	53–91 mequiv./day 2120–3640 mg/day	—
Sodium	140 ± 7 mequiv./l 322 ± 17 mg/100 ml	—	40–156 mequiv./day 920–3590 mg/day	—
Lithium†	0.01 p.p.m. 1 μg/100 ml	—	—	—
Iron†				
males	125 μg/100 ml	iron binding cap:	up to 1 mg/day	—
females	90 μg/100 ml	300 μg/100 ml		
Copper†				
males	105 μg/100 ml	W 92 μg/100 ml	up to 50 μg/day	—
females	114 μg/100 ml	W 97 μg/100 ml		
Zinc	120 μg/100 ml	W 880 μg/100 ml P 300 μg/100 ml	300–600 μg/day	—
Lead	—	W 20–30 μg/100 ml	10–75 μg/day†	—
Arsenic†	3.5–7.2 μg/100 ml	W 6–20 μg/100 ml	< 100 μg/day	—
Boron†	—	—	—	tissue 0.5–1 μg/g
Beryllium†	—	—	—	liver 0.012 μg/100 g
Cadmium	—	—	2–40 μg/day	kidney ash 2–6 mg/g
Chromium†	4.7 μg/100 ml	Cells 2.3 μg/100 ml	—	—
Manganese†	1.3 μg/100 ml	W 2.4 μg/100 ml	0.8 μg/day	—
Nickel†	2.2 μg/100 ml	—	30 μg/day	—
Cobalt	—	W 1–12 (av. 4.3*) μg 100 ml P 1.2 μg/100 ml†	1–10 μg/day	—
Strontium†	—	—	—	bone 0.01–0.25 %

* W = Whole blood; P = Plasma.
† Taken from Christian and Feldman.[8]

operation of changing flames, most clinical laboratories standardize on air acetylene.

Many samples are liquid, making them amenable to the simplest possible pre-treatment—dilution with a spectroscopic buffer. This is the only treatment needed for determining the electrolyte metals—calcium, magnesium, sodium and potassium—in serum or urine. Where such simplified methods are developed, however, a careful appraisal and understanding of the interference effects are necessary.

The presence of protein in serum or whole blood sometimes leads to difficulties. Either the elements may be bound to the protein, leading to some kind of interference effect, or, if the low concentration of the element to be determined does not allow dilution of the sample, clogging of the nebulizer or burner may occur. Deproteinization is then necessary. This step in its simplest form is described on p. 169 under the determination of copper and zinc in serum.

Where sensitivity is a problem, even after deproteinization, solvent extraction can be applied (with or without deproteinization, depending on the element) or if all else fails, wet- or dry-ashing. The latter is also used for solid samples, and usually a useful degree of concentration can be attained, with the possibility of a further sensitivity factor if solvent extraction is used as well.

The earliest workers in atomic absorption considered clinical analyses. Willis in particular, after a series of individual papers[436–43] summarized the method, with reference to biological samples.[444] Cooke and Price[91] published methods for determining sodium, potassium, calcium, magnesium and some trace elements in clinical and biological samples. Dawson and Heaton[119] gave an introduction to the use of spectrochemical analysis, including flame emission and atomic absorption, for clinical materials, with general preparation methods for blood, urine, food and faeces, tissues and bone.

Calcium and magnesium

The relative concentrations and sensitivities of these two elements in serum make it convenient to determine them together in the same dilution, although interfering constituents such as phosphate and protein do not have the same effect in the air acetylene flame.

Several methods have been given in which the serum has been deproteinized before analysis for calcium and magnesium. Willis[439] obtained results by simple dilution with water (1 + 9 or 1 + 24), dilution in the same ratio with sodium EDTA solution (1%) and also after deproteinization with trichloracetic acid, finding them comparable. Gimblet et al.[154] also found trichloracetic acid deproteinization and EDTA dilution comparable but Pybus[313] deproteinized with perchloric acid, adding strontium to remove interferences from phosphate and other anions.

Zettner and Seligson[461] developed a single diluent, containing lanthanum chloride, butanol and octanol, for the determination of calcium, in order to overcome phosphate interference and to improve the response in the flame. Trudeau and Freier[402] simply diluted the serum 1 + 49 in 0.1% lanthanum solution (as $LaCl_3$). More recently, exceptional accuracy was claimed by Pybus et al.[315] using a double channel atomic absorption spectrometer with strontium as internal standard reference element. The serum was diluted 1:50 in a solution which contained 50 mM-hydrochloric acid, 10 mM-lanthanum chloride and 0.12 mM-strontium chloride.

Most workers determining calcium in serum have reported success with urine samples, simply by diluting with lanthanum chloride solution.

The determination of magnesium, in both serum and urine, has been shown to be largely uninfluenced by the presence of phosphate or protein. Dawson and Heaton[119a] diluted samples of serum, plasma, urine, and ashed materials with 0.1 M-hydrochloric acid.

The need for special diluents for calcium determinations suggests a stronger bond between calcium and protein in the serum. In investigating the atomic absorption method for calcium, Cooke and Price[91] laid down that the correct figures must be obtained, whatever the dilution factor employed and in spite of small variations in flame conditions. Phosphate and sulphate both tend to suppress the calcium absorption, as do low concentrations (< 60 mg %) of protein. However, as the protein concentration increases, there is an enhancement in absorption, due to the more reducing nature of the resulting flame. Dilution with water or hydrochloric acid gives the correct result only when the dilution factor is such that the effect of protein counterbalances that of the interfering ions, and thus cannot be relied upon as a generally applicable method. The use of lanthanum as a releasing agent does not appear to fulfil the conditions laid down above, but disodium EDTA solution does.

The releasing effect of disodium EDTA solution rises to a plateau when its concentration is about 0.75 %. Recoveries of added calcium are also good showing that serum-calcium and non-serum-calcium respond in the same way.

Calcium and Magnesium in Serum. Details of the method, which can be used for both calcium and magnesium determinations are as follows:

> Dilute a small volume (e.g. 0.4 ml) of serum with 24 × its volume of disodium EDTA solution, 0.75 %. Aspirate into a just luminous air acetylene flame. Prepare standards containing 0–15 mg/100 ml of calcium and 0–4 mg/100 ml of magnesium and dilute 1 + 24 with 0.75 % disodium EDTA solution.

The accuracy of this method should be in the order of 1–2 % of the calcium and magnesium contents. For some requirements calcium has to be determined with a higher degree of accuracy. Depending on the instrument available, it may then be necessary to use the difference method of calibration, covering the range 9–11 mg /100 ml.

Calcium in Urine. Calcium is determined in urine by the following procedure:

> Dilute the urine so that its calcium concentration lies between 5 and 20 mg/l, including 1 ml of 5 % lanthanum solution (as $LaCl_3$) per 5 ml of final solution. Prepare standards containing 0–20 mg/l of calcium in 1 % lanthanum solution.
> If the results are required in milli-equivalents of calcium per litre, multiply by the factor 0.0499.

Magnesium in Serum. If magnesium and not calcium is required in serum the following simple dilution procedure is used:

> Pipette 0.1 ml of serum into a dry container. Add 5.0 ml of 0.1 M-hydrochloric acid and mix. Aspirate into a just-luminous air acetylene flame. Prepare standards containing 0–1 mg/l of magnesium in 0.1 M-hydrochloric acid. Multiply the concentration of magnesium found by the dilution factor (51). Milli-equivalents per litre = mg/l × 0.0823.

Magnesium in Urine. Dilute the specimen with water so that its magnesium concentration lies between 0.5 and 2 mg/l. The dilution factor will be between 25 and 500. Aspirate into a lean air acetylene flame and compare with standards containing 0–2 mg/l of magnesium as magnesium chloride. Multiply the results obtained by the dilution factor and by 0.0823 to convert to milli-equivalents of magnesium per litre.

Calcium and Magnesium in Solids and Other Specimens. Removal of organic matter by wet or dry oxidation is again the surest approach. Comparable results should be obtained by the three basic methods using dry-ashing, digestion with nitric–perchloric acid mixture and extraction of the wanted metals from dried tissue with 0.5 M-nitric acid. However, it is usually necessary to add releasing agents to the final solution to remove interference from phosphate, nitrate, perchlorate and protein.

Most other methods given for the examination of solid specimens, e.g. skeletal muscle, tissue, hair, faeces etc., are simply variations of this or the wet- and dry-ashing procedures already given in the section on plant materials.

Dry-ashing is also the most suitable method for preparing samples of milk. The residue from 1 g of milk should be dissolved in hydrochloric acid, and diluted to 200 ml so that the final solution also contains 1 % of lanthanum.

Sodium and potassium

The fact that few authors have discussed the determination of these alkali metals in serum or urine in any detail since Willis[440] testifies to the reliability of the method. These elements are often determined by flame emission but the principal advantage of the absorption method is that interference on sodium by the calcium–OH band emission is readily overcome with instruments having modulated sources. The same preparation can be used for either technique.

Sodium and potassium mutually interfere through the shifts which they induce in each others' ionization equilibria. However, as potassium is usually present in lower concentrations in both serum and urine these effects are manifest as a considerable enhancement in potassium response (as compared with a solution containing no sodium) but a very small enhancement of the sodium. Response variations which affect analytical accuracy are countered either by adding the correct amount of the second element to the standards used for determining the first, or by ensuring that an excess of the second element is present in both samples and standards for the first, particularly if the concentration of the second element is variable.

Serum. The normal serum electrolyte levels fall within fairly restricted ranges and the variation of sodium within its range produces a negligible variation of the potassium response. It is therefore necessary, in determining potassium, simply to ensure that the sodium concentration in the standards is comparable to that of the water-diluted samples. A dilution factor of 100 × is suitable. When sodium is determined a greater dilution factor is necessary (250–500 × even with

TABLE 18. Electrolytes in serum

	Ca + Mg	Mg	Na	K
Diluent (sample)	0.075 % $EDTANa_2$	0.1 M-HCl	water	water
Dilution factor	25	50	250–500	100
Diluent (standard)	as sample	as sample	water	35 p.p.m. sodium
Standard range mg/l	Ca 0–6; Mg 0–1.6	0–1	0–20	0–2.5
Air acetylene flame	just luminous	just luminous	lean	lean
Flame path length	10 cm	10 cm	1 cm	10 cm

the sensitivity reduced by burner rotation) and at such dilutions. the potassium concentration is so low that its effect on the sodium response is negligibly small. Dilutions of the sample are made with water, and the standards do not need to contain potassium.

Urine. Because of the possible wide variations in concentration, an excess of one element should be added to the samples and standards used for determining the other. In practice this is necessary when potassium is being determined, but as very low sodium concentrations rarely occur, water can be used as the diluent in this case.

Tables 18 and 19 summarize the preparation of serum and urine samples for determining the four electrolyte elements.

TABLE 19. Electrolytes in urine

	Ca	Mg	Na	K
Diluent (sample)	1.0% La^{3+}	water or 0.1 M-HCl or 1% La^{3+}	water	1000 p.p.m. sodium
Dilution factor	5–50	25–500	200–1000	200–1000
Diluent (standard)	as sample	as sample	water	as sample
Standard range mg/l	0–20	0–2	0–5	0–2.5
Air acetylene flame	just luminous	just luminous	lean	lean
Flame path length	10 cm	10 cm	1 cm	1 cm

Other Materials. When sodium and potassium are to be determined together in the same diluted sample, it is necessary to add another alkali element as ionization buffer.

Because of its low ionization potential, caesium is the best, and its use was described by Sanui and Pace.[344] Alternatively rubidium or lithium—the latter in higher concentrations—may be used, and lithium salts specially free from sodium and potassium are now available for this purpose.

Sodium and potassium may be determined in solid materials, after wet- or dry-ashing or acid extraction by one of the procedures already described for calcium and magnesium.

Lithium

Normal levels of lithium are extremely low, being about 0.01 p.p.m. in serum.[8] Lithium therapy is used in the treatment of manic depressive illnesses and the progress of the treatment is monitored, ensuring that toxic limits are not exceeded, by analysing blood and urine. In these cases, serum lithium levels of about 7 p.p.m. are maintained.

Lehmann[233] diluted serum 1:20 with 0.1 M-hydrochloric acid, obtaining mean recoveries of 101%, though Hansen[162] and Bowman[69] both used 1:10 dilutions of serum in water and the former a 1:50 dilution of urine, also in water. Pybus[314] too used water as the diluent for serum except where high lithium levels were encountered, when sodium and potassium were also added. Blijenberg and Leijnse[61] first deproteinized with 96% ethanol.

Deproteinization, when carried out with trichloracetic acid, ensures that lithium is displaced completely from the protein, and for this reason is probably the better method. Details are:

Into a dry centrifuge tube, pipette 0.5 ml of serum, 2.0 ml of water and 2.5 ml of 10% trichloracetic acid solution. Mix well and centrifuge at 3000 rev./min for ten minutes. Aspirate the supernatant fluid into a lean air acetylene flame. Prepare standards with up to 2 μg/ml of lithium in a solution also containing 5% of trichloracetic acid and 322 μg/ml of sodium as sodium chloride.

Iron

Serum. Abnormal serum iron concentrations may be found in various types of anaemia, in the presence of malignancy, and in pregnancy. The normal levels of iron in serum (~ 1 p.p.m.) are close to the analytical sensitivity value, and therefore means usually have to be found, either of concentrating the iron, or of improving the sensitivity. The spraying of undiluted serum may cause instability of the flame and blockage of the burner and nebulizer system. Zettner et al.[462] used solvent extraction of the bathophenanthroline derivative with no apparent advantage over a colorimetric ending, and Olson and Hamlin[287] deproteinized with 20% trichloracetic acid. Deproteinization probably forms the basis of the most accurate method, particularly if the response of iron is improved by addition of acetone to the analysis solution. Such a procedure, designed to achieve the highest practicable ratio of sample to final volume, is as follows:

Stand 2 ml of serum with 0.1 ml of 4M-hydrochloric acid for one hour then add 1 ml of 20% trichloracetic acid, mix well and centrifuge. Transfer an aliquot of the supernatant of exactly half the total volume (1.55 ml) to a dry tube, add 0.5 ml of acetone and aspirate into a lean air acetylene flame. Prepare standards each containing 2 ml of iron standard solutions (0–3 p.p.m. Fe), 0.1 ml of 4M-hydrochloric acid, 1 ml of 20% trichloracetic acid and 1 ml of acetone. These standards are equivalent to 0–300μg of iron per 100 ml of the original serum.

If highly stable flame and instrumental conditions can be achieved, so that scale expansion factors of 10× or more are possible, comparative serum iron determinations can be made simply by dilution of the serum 1+2 or 1+4 with water, comparing with aqueous standards.

In addition to serum iron, the iron-binding capacity may be determined by a simple extension of the method.[287] A sample of serum is saturated with ferric chloride solution, and the excess is removed with magnesium carbonate. After centrifugation the supernatant is deproteinized with trichloracetic acid, again centrifuged, and the final supernatant aspirated as for serum iron, giving the total iron-binding capacity.

Whole Blood. Iron determinations in diluted whole blood are of interest as a simple method of determining blood losses in operation and menstruation, the swabs being soaked in a known volume of water which is aspirated directly and compared with aqueous iron standards. An important determination, upon which depends the accuracy of the cyanmethaemoglobin spectrometric method for the measurement of haemoglobin levels in blood, is that of iron in haemoglobin itself. Using fresh red cells lysed with toluene, Zettner and Mensch[460] established a value somewhat at variance with that accepted by the International Committee for Standardization in Haematology in 1967. van Assendelft et al.[408] have since suggested that both the presence of inorganic substances and the handling of blood samples influence the results, illustrating the point by comparing their own with Zettner's procedure, and results obtained with different instruments.

Urine. Low iron concentrations may indicate excessive iron storage or haemolysis. Low and normal iron in urine can be determined by extraction with ammonium pyrrolidine dithiocarbamate into methyl isobutyl ketone. High iron concentrations may be determinable by direct aspiration.

Copper and zinc

Blood. Because of the good sensitivity of these two elements, they can be determined at normal levels in serum, plasma and whole blood without concentration by extraction. A raised copper level is present in tissues and in serum in Wilson's disease. Zinc deficiency often accompanies slow healing of wounds.

A wet-ashing technique was developed by Delves[122] to handle small samples of blood from children with pica. The digestion mixture used was perchloric, nitric and sulphuric acids, and after evaporation to dryness the residue was dissolved again in hydrochloric acid.

Several workers, including Parker et al.[292] have indicated that copper and zinc can be determined in serum either after simple dilution or after deproteinization. Olson and Hamlin[286] separated the metals from the protein by heating with 20% trichloracetic acid. Deproteinization is undoubtedly preferable in a procedure intended for general applicability as it ensures that no protein-bound metals remain, and also a cleaner solution is thereby presented to the atomic absorption spectrometer. A suitable method for both copper and zinc is:

> Pipette 1 ml of serum and 1 ml of 10% trichloracetic acid into a dry centrifuge tube. Mix carefully and centrifuge at 3000 rev./min for ten minutes. Aspirate the supernatant into a lean air acetylene or air propane flame. Prepare standards containing 0–1 mg/l of copper and zinc in 5% trichloracetic acid. Multiply the results obtained by the dilution factor 2.
>
> With a sensitive instrument, a 1 + 4 dilution of the serum with water is often adequate. Comparison is then made with aqueous standards.

Urine. Excretion of zinc is usually at a higher level than that of copper, and accounts for zinc being a normal contaminant of domestic sewage. Low copper concentrations in urine, present in cases of neoplasm, may be determined after extraction with ammonium pyrrolidine dithiocarbamate into methyl isobutyl ketone. High levels of copper and zinc can usually be determined by direct aspiration.

Other Materials. Solid specimens require wet- or dry-ashing, followed by further concentration, if necessary, by solvent extraction. Backer[41] claimed that chloric acid digestion was simpler than perchloric–nitric acid mixtures, and is suitable for hair and brain tissue samples.

Blomfield et al.[62] used atomic absorption to estimate patients' uptake of copper and zinc during regular haemodialysis. The total copper and zinc of red cells and free copper and zinc of plasma and dialysis fluids were determined at all points during the process as well as subsequent blood levels. Copper plumbing and zinc oxide plasters were found to be the major source of contamination by these elements.

Lead

There is rapidly increasing public concern about the degree of pollution of the environment by lead from various sources[77] and particularly from the combustion of lead anti-knock agents in gasoline. Many thousands of workers in industry are directly exposed to the dangers of lead poisoning (metals, batteries, paints etc.) and also children, through the paint and metal of their toys and surroundings. The determination of lead as a routine screening test will therefore assume greater importance in coming years in industrial hygiene, public health, paediatric and clinical laboratories.

The normal blood and excretory levels (Table 17) are low and, when raised, indicate lead poisoning. The action of lead is insidious as it tends to accumulate in the body, causing damage to the brain and central nervous system.

Blood. In the blood lead concentrates in the red cells, and the generally accepted danger level is 80 mg/100 ml. If a worker's blood lead concentration approaches this level, he should be removed from the source of ingestion.

Because of the low normal level of lead in blood, useful dilution factors cannot be employed in sample preparation, and thus separation of lead from the protein is a necessity. Deproteinization with 5% trichloracetic acid was the basis of a method given by Lehnert and Schaller[235] but 10 ml of blood were required, which would now be considered to be unnecessarily extravagant. Selander and Cramer[354] compared two methods for the preparation of heparinized blood samples. In the first place the samples were digested with nitric–sulphuric acid mixture, then the pH was adjusted to 3 and the lead was extracted with ammonium pyrrolidine dithiocarbamate into methyl isobutyl ketone. In the second method, the extraction was preceded by deproteinization with trichloracetic acid. The first method was used because, as the authors pointed out, the usual way of assessing these methods is to evaluate the recovery of lead added to the samples in known quantity, but it is not certain that lead present in the biological material will react in the same way as that added. The wet oxidation overcomes this difficulty. However, they found that good agreement was obtained on a number of samples from normal, occupationally exposed and lead poisoned patients with the two methods.

Farrelly and Pybus[135] took advantage of the concentration of lead into the red cells, but added formamide before extracting the lead with ammonium pyrrolidine dithiocarbamate in order to break the emulsions formed. Unfortunately formamide has been shown to be a teratogenic substance, and should not be handled by female laboratory assistants. The following procedure is harmless from this point of view, but ensures the best sensitivity to be gained in sample preparation:

> Pipette 5 ml of heparinized whole blood into a graduated 10 ml stoppered centrifuge tube. Centrifuge for 20 minutes at 3000 rev./min then remove and discard the plasma. Note the volume of the packed red cells. Add 1 drop of saponin/triton solution (5 ml of Triton X plus 5 g of saponin in 25 ml of water) and mix. Into the mixture, pipette first 2 ml of 2% ammonium pyrrolidine dithiocarbamate solution, shake for 30 seconds, and then add 3·0 ml of methyl isobutyl ketone. Stopper the tube, shake for 1 minute then centrifuge at 3000 rev./min for five minutes. Aspirate the top layer into a lean air acetylene flame. Prepare standards by taking 2 ml of lead standard solutions

containing 0–2 μg of lead per ml through the procedure. Read the lead content of the sample extracts in μg, then lead content of red cells in μg/100 ml = μg in extract × 100/v where v is the volume of the red cells.

The method can be 'miniaturized' and applied to whole blood samples, and may then be reliably used in industrial hygiene and paediatrics:

Pipette 1 ml of well mixed heparinized blood into a stoppered centrifuge tube. Add 1 drop of saponin/triton solution (see above), mix thoroughly and add 1 ml of 2% ammonium pyrrolidine dithiocarbamate solution. Shake the tube for 30 seconds then add by pipette 1 ml of methyl isobutyl ketone and shake for 1 minute. Decant the mixture into a small-bore thick glass test-tube and centrifuge for 10 minutes at 3000 rev./min. Aspirate the top layer into a lean air acetylene flame. Prepare standards by taking 1 ml of lead standards containing 0–1 μg of lead per ml through the procedure. Read the lead content of the sample extracts in μg, multiplying the result by 100 to express it in μg of lead per 100 ml of whole blood.

Many attempts have been made to increase the sensitivity of the lead determination to the point where normal levels are determinable in a drop of blood obtained from the prick of a finger or ear-lobe—e.g., a 20 μl sample. This precludes any dilution or even nebulization. 'Sampling boats' and long tube flame cells (see Chapter 3) have been used in expert laboratories, but the most practical approach, giving adequate accuracy for mass screening appears to be that of Delves[123] who described a total combustion method in which 10 μl of blood were partially oxidized with hydrogen peroxide, then vaporized from a nickel micro-crucible (10 mm × 5 mm) placed beneath a 10 cm nickel absorption tube. The sensitivity of the method is said to be 10^{-10} g of lead. The apparatus is available commercially.

The main problem with such micro-sampling methods as applied to industrial screening is that contamination levels arising from lead deposited on the surface of the skin can be comparable to the levels actually being determined. Sampling by venipucture does not incur this difficulty and a 1 ml sample is as readily withdrawn as 10 μl. It is also essential that all apparatus used, whichever method is employed, be kept well away from all possible sources of contamination, and that glassware be soaked overnight in 1 + 1 nitric acid and repeatedly rinsed in deionized water.

Urine. Lead is determined in urine in order to follow the progress of therapy, which usually consists of the administration of a lead-chelating reagent such as EDTA or penicillamide. When the former is used, however, excreted lead is so strongly bound that the standard solvent extraction procedure using ammonium pyrrolidine dithiocarbamate is ineffective.

Selander et al.[353] used wet-ashing before extraction, and for lead concentrations below 5 μg/ml, which could not be measured by direct aspiration of the sample. Roosels and Vanderkeel[334] added calcium to displace lead from its EDTA complex so that the lead could be extracted with dithizone. This leads somewhat inconveniently to a chloroform solution.

The problem has also been solved by co-precipitation methods. Kopito[218] precipitated lead—with other polyvalent ions—from ammoniacal samples with bismuth nitrate. Zurlo et al.[464] co-precipitated with thorium in the presence of copper which displaces lead from its EDTA complex. The method is made quite simple:

> To 12.5 ml of the sample (brought to pH 5–6 with ammonia) in a centrifuge tube add 0.5 ml of thorium–copper solution (containing 2 g of thorium nitrate, 2g of copper sulphate crystals and 1 drop of concentrated hydrochloric acid in 100 ml). Mix and centrifuge at 3000 rev./min for 5 minutes. Discard the supernatant and dissolve the residue in 0.5 ml of 50% hydrochloric acid and 2.0 ml of water. Aspirate into an air propane or lean air acetylene flame. Compare with simple aqueous standards containing 0–4 mg/l of lead as lead nitrate, acidified with a drop of hydrochloric acid per 100 ml.

Where therapy is known to have been by way of penicillamide, extraction of the lead with ammonium pyrrolidine dithiocarbamate is satisfactory:

> Adjust the pH of a 50 ml sample aliquot in a 100 ml separating funnel to 2.5 ± 0.1 with hydrochloric acid, add 5 ml of 1% ammonium pyrrolidine dithiocarbamate solution and 6 ml of methyl isobutyl ketone. Shake for 10 minutes, allow to stand for 5 minutes, discard the lower (aqueous) layer, and collect the organic layer and any emulsion in a 10 ml centrifuge tube. Centrifuge for 10 minutes at 3000 rev./min and aspirate the organic layer into an air propane or lean air acetylene flame. Compare with standards prepared by taking 50 ml aliquots of lead solution containing 0–1 μg/ml of lead through the procedure. These standards are equivalent to 0–1 μg of lead per ml of the urine sample.

Other elements

While the foregoing elements are the ones most frequently determined in clinical laboratories, they are by no means the only ones. Not only are some elements administered in therapy or as markers, but as the determination of trace metals becomes easier and more sensitive, 'normal' values become more readily established and abnormal values related to certain conditions or diseases. In principle, any metallic element can be determined in biological material provided that sufficient sample is available from which an adequate quantity of the required metal can be separated.

Arsenic. The normal level of arsenic in urine is thought to be not more than 0.1 p.p.m., and 1 p.p.m. is considered to be evidence of harmful exposure. Poisoning cannot be inferred from the presence of arsenic in the stomach contents as it may not have been absorbed.

Few direct determinations of arsenic by atomic absorption have been reported. Samples cannot be prepared by dry-ashing as arsenic is lost by evaporation at temperatures well below 500°C. Even with nitric–sulphuric–perchloric acid digestion some arsenic is lost[357] but this is prevented by addition of molybdenum VI. Arsenic can be extracted by ammonium pyrrolidine dithiocarbamate into methyl isobutyl ketone at pH 0–4.

Special flames and conditions for improving the sensitivity of arsenic are given in Appendix 1.

A gas-sampling technique for increasing the amount of arsenic actually reaching the flame atomizer was described by Holak.[174] The sample is placed in a conventional arsine generator fitted with a calcium chloride drying tube. This in turn is connected to a U-tube in a liquid nitrogen trap. The arsine is first collected in the

U-tube, which is then removed from the liquid nitrogen and connected to the uptake capillary of the atomic absorption spectrometer. All the arsenic then passes to the flame. As no other materials are present (except possibly stibine) the flame conditions described in Appendix 1, giving best sensitivity for arsenic can be used. The amount of arsenic present is proportional to the area beneath the recorder output trace.

Boron. Bader and Brandenburger[42] stated that sub-toxic levels of boron in serum and urine could be determined by direct aspiration into a nitrous oxide acetylene flame. For lower concentrations, or for tissue and bone, a wet oxidation procedure must be used as dry-ashing leads to severe boron losses.

Beryllium. Beryllium is an extremely toxic metal, causing berylliosis. It also affects bone calcification mechanisms and displaces magnesium from, and inhibits, magnesium dependent enzymes. Normal treatments for beryllium poisoning aim at rendering the beryllium insoluble and retaining it in the tissues.[163]

Bokowski[65] detected 0.002 p.p.m. of beryllium in urine using the nitrous oxide acetylene flame, while Butler[79] separated beryllium from prepared solid specimens and urines by two liquid ion exchange procedures, using di-2-ethyl-hexyl phosphoric acid and triiso-octylamine respectively. The latter gave 90% recovery of 1–12 μg of beryllium, was quicker, and did not require a critical pH adjustment.

Cadmium. Because of the high volatility of this metal workers exposed to fumes may exhibit chronic poisoning. Such workers show much increased urinary excretion levels (40–400 μg/l instead of 2–20 μg/l[365]) though continued exposure may result in accumulation in the kidney, displacing zinc to give hepatic zinc deficiency.

Urine analysis appears to be a good method for screening purposes, and this was described by Lehnert et al.[234] Significant levels should be determinable by direct aspiration. Otherwise cadmium is determinable with exactly the same procedures as described for lead—e.g. by extraction from urine with ammonium pyrrolidine dithiocarbamate.

Chromium. Chromium is used as a faecal marker, and there is concentration with time, in the spleen. Plasma chromium concentration also appears to be related to glucose tolerance and may be of some use in diabetes diagnoses.

Williams et al.[435] determined chromic oxide in faecal material after an ashing procedure, and Feldman et al.[139] dry-ashed serum, plasma, whole blood, urine and diet at 550°C as well as using a wet decomposition method.

Gold. Gold is not believed to be an essential element, but may be accumulated as an environmental contaminant. Gold complexes, e.g. sodium aurothiomalate are injected in the treatment of rheumatoid arthritis. The element may therefore have to be determined in the injection material itself[363] (also called myochrysine) and in patients' sera during therapy to avoid overdosage which may cause unpleasant side effects. The change in serum gold levels with time following such an overdose was determined by Christian.[87] Lorber et al.[239] used the method of standard additions to measure gold levels in serum, urine and synovial fluid, the diluent for serum being sodium dodecylsulphate solution.

Manganese. Although manganese levels are normally very low its distribution appears to be of some interest. Mahoney et al.[243] determined manganese in serum by the method of standard additions and earlier Koirtyohann and Feldman[215] used a long path absorption tube for tissue ash manganese determinations, comparing with a synthetic ash matrix as a basis for standards.

Mercury. Studies in environmental pollution as well as industrial hygiene have prompted considerable interest in this element, and the development of simple flameless atomic absorption measurement techniques of very great sensitivity have facilitated its determination at low levels.

Workers handling mercury metal or compounds face the greatest risk of poisoning, and the analysis of urine provides a satisfactory screening test. The severity of symptoms, however, is not always in direct relation to the level found. Atomic absorption methods do not distinguish between different forms of ingested mercury, and therefore, as well as poisoning, diuretics and mercurial antiseptics are possible sources.

Loss of mercury by volatilization during sample preparation is the main problem in this analysis. Neither dry-ashing nor hot digestion can be considered. Lindstedt[238] described a flameless method for determining mercury in urine in which the organic material was oxidized by standing 1 ml samples in the cold overnight with 1.5 ml of 6% potassium permanganate solution and 0.2 ml of sulphuric acid. The sulphuric acid was added first and the mixture cooled before adding the potassium permanganate. The following day the excess permanganate was reduced with 0.3 ml of 20% hydroxylamine hydrochloride solution. This solution was then treated by a method similar to that described in Appendix 1 under Mercury.

Mercury can be determined in tissue samples after oxidation at a somewhat higher temperature. Uthe et al.[406] for example, found 50°–60°C to be adequate for complete recovery of mercury from fish tissue. A temperature of up to 70°C is permissible for the digestion of this type of sample, and the following procedure is safe and reasonably fast:

> Place 0.4–0.8 g of homogenized sample in a weighed 150 ml conical flask with a B24 socket neck. Reweigh and cover with a watchglass. Pipette 5 ml of sulphuric acid into the flask and place in a water bath at 70° C for one hour, at the end of which the solution should be homogeneous though highly coloured and perhaps a little cloudy. Cool in an ice bath, add carefully 50 ml of 6% potassium permanganate solution, and replace in the water bath for two hours. Cool to room temperature and add, by pipette, 15 ml of 20% hydroxylamine chloride. Prepare two reagent blanks by taking 2 ml of water through the same procedure.
>
> The measurement is made with the apparatus shown in Fig. 40 (see details p. 197). Immediately before connecting the socket of the flask to the Drechsel bottle head of the measuring apparatus, introduce 2 ml of 10% stannous chloride solution, then switch on the pump. The absorbance value increases to a constant level in 1–2 minutes. When this value has been recorded, disconnect the ground glass joint and continue the aeration until the absorbance returns to its minimum value. The equipment is then ready to receive the next sample. Calibration solutions containing 0–3 ml of mercuric sulphate solution, 0.1 mg Hg/l (≡ 0–300 ng Hg) are taken through the measurement sequence after reduction with stannous chloride solution as in the preceding paragraph. The standards should be prepared immediately before analysis. Plot a graph of ng Hg against absorbance, subtracting the absorbance of the zero standard from the other standards. Subtract the reagent blank absorbance from sample absorbance readings and read off ng Hg from calibration graph. The reagent blank should not be more than 30 ng of mercury, and if it is, the reagents and glassware should be checked.

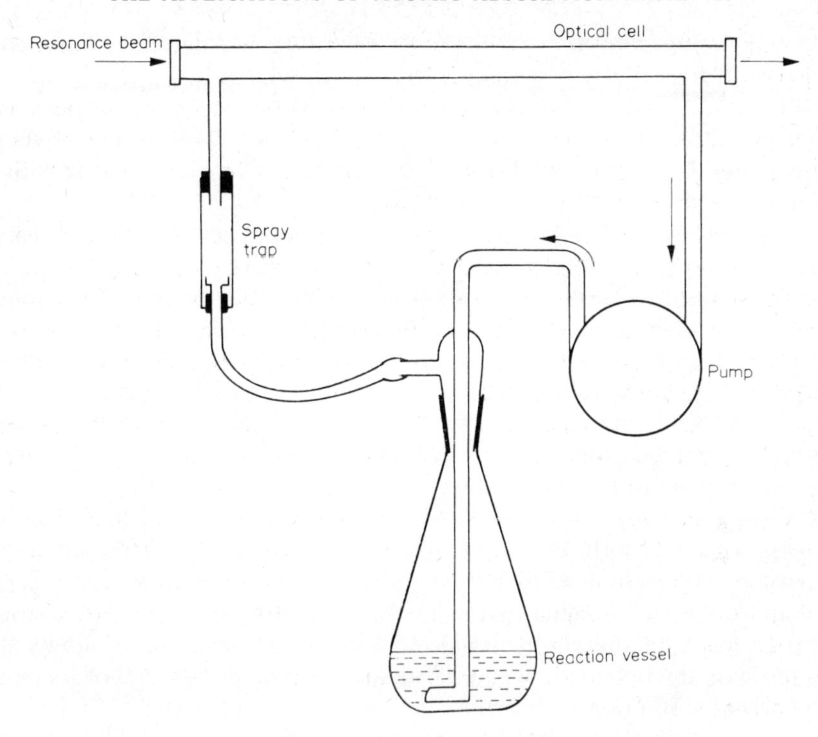

FIG. 40. Apparatus for cold vapour determination of mercury.

The properly digested mixture may contain suspended hydrated oxides of manganese. These should dissolve completely giving a colourless or slightly opalescent solution when the reductant is added. Digestion with sulphuric acid and potassium permanganate is said to release mercury in all forms, including that present as methylmercury.[288]

Conventional methods involving the extraction of mercury with ammonium pyrrolidine dithiocarbamate were described by Willis[443] and Berman[55] for urine and blood respectively.

Nickel and Cobalt. These elements have both been measured directly in urine[371,443] but, if the levels are too low, can readily be determined after extraction with ammonium pyrrolidine dithiocarbamate into methyl isobutyl ketone. A number of heavy metals are extracted together at pH 2.8.

Other materials generally require wet-ashing. A further separation can be made after the wet-ashing step by making the solution alkaline and extracting the nickel dimethylglyoxime complex into chloroform. The nickel must then be back-extracted into 0.5 M-hydrochloric acid.

Sullivan et al.[380] analysed post-mortem heart tissues from 'beer drinkers' myocardiopathy patients after the samples were wet-ashed with nitric acid. An abnormally high cobalt level—confirmed by neutron activation analysis—suggested that cobalt salts, added to some American beers, were to blame.

Strontium. The similarity in chemical behaviour between strontium and calcium causes strontium to be absorbed in bone—a cause for serious concern because of the increase in radiostrontium from nuclear fallout. This similarity can be utilized in chemical preparation and concentration steps by co-precipitating strontium and calcium and then determining strontium after redissolution. Curnow et al.,[99] analysing serum, redissolved in the lanthanum-containing diluent, while Montford and Cribbs[275], analysing urine, co-precipitated with lanthanum, and redissolved in hydrochloric acid. Tompsett[397] used Dowex 50W cationic exchange resin to concentrate strontium after dry-ashing samples of urine and faeces.

Thallium. There is some danger of thallium poisoning as in some countries this metal is used in rat poisons. Its action in the body is similar to that of lead, but in atomic absorption it is more sensitive. Curry et al.[100] described methods for determining thallium in biological material. Blood, tissue and stomach contents were digested with sulphuric and nitric acids. The thallium was then converted to bromide at pH 3–4 (adjusted with ammonia) and extracted into water-saturated methyl isobutyl ketone. This specific extraction method for thallium can be used for both atomic absorption and flame emission measurements. It is suggested that when atomic absorption is used, specificity is not required and thallium can be extracted with sodium diethyl dithiocarbamate instead of as bromide. Tissue, blood and urine were treated by this method, digesting as above, but adjusting the pH to 5–6 with sodium hydroxide solution. Absolute detection limits were 40 ng for the extraction method and 200 ng for a direct aqueous method, requiring 1 ml of prepared solution. A detection limit of <2 ng was reported when using the tantalum boat technique, which required 50 µl of urine or prepared sample.

Other topics

The use of atomic absorption analysis in forensic medicine was discussed by Brandenburger.[72] The determination of small amounts of silver, copper and mercury was described.

The occurrence and function of trace metals in biological systems, particularly metallo–enzymes and metal–enzyme complexes, were discussed by Vallee.[407] Atomic absorption was used in the analysis of these materials, and in the investigation of some of the reaction mechanisms. Fractions of zanthine oxidase were also analysed for molybdenum, iron and copper by the method of standard additions after dry-ashing.[336]

The analysis of human hair as a diagnostic aid was investigated by Harrison et al.[164] Methods of cleaning and preparation were described in some detail, and copper, iron, magnesium and zinc were determined after digestion with nitric acid and fuming with perchloric acid.

Phospholipid sodium, potassium, calcium, magnesium and manganese contents were determined by Montford and Cribbs[276] after dissolution of the samples in isopentyl acetate. To suppress the interference of calcium, an ethanolic solution of lanthanum was added. Results given by a 'direct' method and a standard additions procedure were in close agreement, indicating that the interferences had been successfully overcome.

Although the tolerance limits in the analysis of the dialysis solutions used in artificial kidney machines are very close, e.g. 1% (or less) relative concentration for the four electrolyte metals, it has been shown that such limits can be attained by the use of difference measurements.[389] Differences between solutions of the following two compositions (all in mg/l)

	1	2
Sodium	1000	1010
Calcium	21.2	21.4
Potassium	13.9	14.0
Magnesium	4.30	4.35

can be expanded to about 5–10% of the full absorbance output scale. The air acetylene flame gives adequate stable response for sodium, potassium and magnesium, but nitrous oxide acetylene may have to be used for calcium in order to increase the absorption signal.

INDIRECT ATOMIC ABSORPTION METHODS

By 'indirect' methods are usually meant those cases where the species being determined is itself not measured, or may not even be measurable. Instead, a metal is measured which is in some way related chemically to the wanted species. A number of different types of indirect atomic absorption methods were listed by Pinta.[296] Some anions or other organic species can be measured by their interfering effects upon the absorbance of some metal. This is perhaps the most obvious example. Precipitation of the species to be determined, followed by atomic absorption of the excess precipitant is another, and 'chemical amplification' such as formation of a heteropoly acid from phosphorus, silicon, vanadium etc. and determination of the molybdenum which is present in a much greater proportion in the resulting molecule, is a third.

Some examples of indirect methods are given here, but many analytical chemists will be able to apply similar principles to their own specific problems.

Use of interfering effects

Oxyacids. Aluminates, phosphates, sulphates, silicates and other oxyacids exert a considerable depressive effect on the atomic emission and absorption of calcium and strontium in air acetylene or air propane flames. The reasons for this have been discussed earlier in this book. Pinta[297] has shown that it is possible, using 20 p.p.m. of calcium as the basis for measurement, to obtain a graphical relationship for concentration of phosphate (P between 4–20 p.p.m.) and sulphate (S between 4 and 10 p.p.m.). A method for determining phosphate in waters, rocks and steel was described also by Singhal et al.[360] making use of the depression of strontium absorption.

Fluorides. In low temperature flames, fluoride depresses the response of magnesium.[67] Fluoride between 2 and 15 p.p.m. can be measured by its effect on the absorbance of 10 p.p.m. of magnesium. Phosphate and sulphate interfere, but the other halides and many metals do not.

Fluoride increases the response of both zirconium and titanium in the nitrous oxide acetylene flame, and this phenomenon has been used by Bond and Willis[68] also to determine fluoride.

Nitrogenous Materials. The same workers showed that ammonia, amines, amino-acids and other organo-nitrogen compounds also increased the response of zirconium. For example, concentrations between 1.7 and 85 p.p.m. of ammonia increase the absorbance of 1350 p.p.m. of zirconium rationally, in the presence of 0.006 M-potassium chloride as ionization buffer.

Other Organic Materials. The depressive effect of protein on calcium has already been discussed. Sugars exert a similar effect on calcium in low temperature flames[88] and concentrations below 10^{-6} M may be related to the absorbance of 25 p.p.m. of calcium. Proteins were determined in a similar procedure. As the concentration of these organic materials increases, the depressive effect reverses into an 'enhancing' effect. The value of such methods is therefore limited to certain concentration ranges.

Determination of Traces of Aluminium and Titanium. The increase in absorbance of iron in a stoichiometric air acetylene flame in the presence of very low concentrations of aluminium and titanium was reported by Ottaway,[289] and this observation was made the basis of a method in which the detection limits of these two elements were 0.02 and 0.01 p.p.m. respectively.

General. 'Interference' methods clearly have a limited application, as the extent of the effect depends critically on flame temperature and flame chemistry. Flame conditions must therefore be stable and reproducible. Furthermore, the effects are rarely specific. They can usually be employed only if the sample solution contains no variable constituent other than the substance to be measured.

Precipitation methods

Sulphates are determined by precipitation with a known quantity of barium, followed by determination of the excess by atomic absorption. Ecrement (see Pinta[296]) determined between 5 and 60 p.p.m. of sulphate, adding 0–100 p.p.m. of barium. An alternative approach would be to determine the precipitated barium by separation of the barium sulphate, and redissolving in ammonium EDTA solution.

Rose and Boltz[335] determined sulphur dioxide by formation and precipitation of lead sulphate.

Chlorides (5–100 p.p.m.) may be precipitated with the addition of 300 p.p.m. of silver.[296] Determination is made either of the remaining silver, or of that precipitated after separation and redissolution in ammonia.

Reducing agents may be determined by precipitation of copper II. Between 0.1 and 5 p.p.m. of dextrose were determined by Potter et al.[303] using this principle.

Chemical amplification with heteropoly acids

Ammonium salts of heteropoly acids are formed in solution between ammonium molybdate and phosphate, silicate, titanate, vanadate, arsenate, niobate and germanate ions. The compounds formed are stoichiometric under given conditions of temperature, pH etc., and are extractable—often selectively—into organic solvents. This principle is, of course, well known, in spectrophotometry. The fact of particular interest in atomic absorption is that the ratio of molybdenum to combining element is very high. Silicon forms the complex $NH_4SiO_4(MoO_3)_{12}$ and the weight ratio of molybdenum to silicon is 41 : 1. If the molybdenum absorbance is measured there is thus a considerable 'amplification' factor over the silicon and this is further increased by the extraction factor and the improvement expected from the use of an organic solvent.

Zaugg and Knox[458] used this principle to determine phosphate, and Kirkbright et al.[209] extended it first to the sequential determination of phosphorus and silicon, and later to niobium[210] and titanium.[211]

Hurfurd and Boltz[180] also discussed the determination of phosphate and silicate and Ramakrishna et al.[317] gave procedures for these elements and arsenic.

Some details and references are summarized in Table 20.

TABLE 20

	Complex extracted	Mo/el. ratio	Extraction pH	Extraction extractant	Ref.
Phosphorus	phosphomolybd.		pH 0.7	isobutyl acetate	317
Arsenic	arsenomolybd.		pH 0.7	(1 + 1) ethyl acetate + butanol	317
Silicon	silicomolybd.	41	pH 3.2	MIBK	317
Phosphorus	phosphomolybd.		0.15 M-HCl	chloroform/butanol	211
Titanium	P–Ti–molybd.		0.15 M-HCl	butanol	211
Titanium	P–Ti–molybd.	24	pH 1	isobutyl acetate	296
Vanadium	P–V–molybd.	20–6	ammonia	(4 + 1) diethyl ether + pentanol	185
Niobium	P–Nb–molybd.	11.3	0.5 M-HCl	butanol	210

Determination of compounds forming extractable metal chelates

In the same way as metals are extracted quantitatively after addition of excess chelating agent, so may small quantities of the chelating agent be extracted after addition of excess of a suitable metal. The extracted metal chelate may then be aspirated to obtain the chelating agent concentration by measuring the metal in atomic absorption, or alternatively the excess metal remaining in the aqueous solution can be measured, giving the chelate by difference.

Dithiocarbamates, 8-hydroxyquinoline and EDTA have all been determined in this way.[296] Extractable complexes with metal chelates are also given by some compounds. Kumamaru[219] determined phthalic and nitric acids by extracting the complexes which they form with the copper I—neocuproin chelate. Sensitivity for phthalic acid was between 4×10^{-6} and 4×10^{-5} M. Collinson and Boltz[90] determined perchloric acid (0.2–6 p.p.m.) with the same chelate. Thiocyanates may be determined by the chloroform-extractable $Cu(py)_2(SCN)_2$ complex with pyridine and iodates[455] by their nitrobenzene-extractable cadmium—1:10 phenanthroline complex. An indirect determination of mercury II by extraction with 1,2′-bipyridyl zinc chelate followed by determination of the zinc was also described.[456] The sensitivity of mercury thus becomes similar to that of zinc, but a number of metals may interfere.

Oxidizing and reducing agents

Making use of the fact that chromium VI can be extracted from acid solution into methyl isobutyl ketone, if part of the chromium is first reduced to Cr III by another compound for example, iodide, the latter can be determined in relation to the chromium actually extracted.[88] Conversely, iron is converted from Fe II to Fe III by iodates and the latter can be determined by extracting the resulting Fe III in presence of hydrochloric acid into ether or another more suitable solvent.

Molecular absorption in flames

Though not strictly within the scope of this book, these methods are mentioned, as the atomic absorption spectrometer is used to make the measurements. There are many instances where molecular absorptions are quoted as interferences in atomic absorption determinations. Fuwa and Vallee[147] used the absorption bands of sulphur dioxide between 200 and 300 nm to determine sulphur in organic compounds with a long tube flame cell and total consumption air hydrogen burner. The radiation source was a hydrogen lamp.

THE USES OF ATOMIC FLUORESCENCE ANALYSIS

The number of known genuine analytical applications of atomic fluorescence spectroscopy is very small though there has been a great deal of valuable information published by research workers on the sensitivity of various elements with different sources of excitation and in different atomizing flames. Much of this has been initiated by West, who discussed[428] the practicality of atomic fluorescence as an analytical technique. He suggested that scatter and quenching effects should not generally limit the sensitivity that can be obtained. The selection of atom reservoir is more significant as a choice may have to be made between one which gives best sensitivity for the test element, and one which achieves complete vaporization of the sample.

Flame atomization

An ideal flame atomizer for atomic fluorescence should give the highest degree of atomization together with the lowest emission background and flame noise. Unfortunately these requirements are often incompatible. Source modulation allows the emission signal to be rejected, but will not overcome the effect of noise associated with this signal. Use has therefore been made of low background hydrogen based flames, but for reasons discussed in the section on atomization (p. 19 et seq.) these do not give an adequate degree of atomization for the so-called refractory elements.

Primary source

Provision for a primary excitation source may be the second largest outlay when atomic fluorescence methods are being considered. The performances of various excitation sources for a number of elements are therefore summarized briefly.

Continuum Source Excitation. Although Winefordner and Vickers[452] first suggested the continuum source in 1964, the range of elements determined with this type of source has not altered appreciably since further work was reported from the same laboratory two years later. Then, Veillon et al.[412] investigated 13 elements using a 150 W xenon arc lamp in conjunction with a phase-sensitive amplifier. Good detection limits were obtained for bismuth, copper, gold, lead, magnesium, silver, thallium and zinc in an oxyhydrogen flame and for lead, magnesium and thallium in an air hydrogen flame. Scattered radiation was a serious interference, and Cresser[94] stated that even when the fluorescence spectrum is scanned, scatter contributes considerably to the background noise. He also suggested that the fluorescence excited by a continuum, in the type of flame necessary to ensure adequate atomization of many elements, is lower in intensity than the simultaneous thermal emission. Cresser and West[97] investigated the fluorescence of thirteen elements in a premixed air acetylene flame excited by a 500 W xenon arc. They found that the interferences were similar to those encountered in emission, though molecular band emission did not interfere. Detection limits for many of the elements including copper, silver, indium, thallium, cadmium, calcium, magnesium, chromium, manganese, cobalt and nickel were better in emission. An account of this work was also given by West.[428]

A continuum source, the 450 W xenon arc, was used by Cotton and Jenkins[92] in the analysis of kerosene for traces of copper, iron and lead. A burner was specially made, packed with strips of corrugated stainless steel, to give flat premixed flames with hydrogen or with a liquid hydrocarbon fuel. With kerosene itself, for example, and with benzene, the flame was non-luminous and its background emission was consequently low. The detection limits were copper 0.004, iron 0.04 and lead 0.06 p.p.m.

Hollow Cathode Lamp Excitation. Hollow cathodes of the high spectral output or boosted output types often give sufficient intensity to excite fluorescence, though very good limits of detection cannot be expected. Nevertheless, West and Williams[429,430] determined silver and magnesium in an air propane flame with detection limits of 0.0009 and 0.0008 p.p.m. respectively. Beryllium was determined by Robinson and Hsu[327] in an oxyacetylene and also a nitrous oxide acetylene flame. Best freedom from the many potential interferences investigated was found in the latter.

Matousek and Sychra[255] reported detection limits of 0.02, 0.01 and 0.003 p.p.m. for iron, cobalt and nickel in an oxygen–argon–hydrogen flame, and 0.005 p.p.m. for gold[257] in a similar, though separated, flame. Sychra et al.[385] also investigated the atomic fluorescence of palladium, finding the best detection limits (0.04 p.p.m.) with an oxygen–argon–hydrogen flame. In the application of atomic fluorescence, also to petroleum products, Sychra and Matousek[384] used a boosted output hollow cathode to determine nickel, with a detection limit of 0.007 p.p.m. The samples, gas oils and petroleum distillation residues, were atomized in a nitrogen-separated air acetylene flame. The nickel line 232.0 nm gave good linearity of calibration from less than 0.01 to about 5 p.p.m. of nickel in solution

Electrodeless Discharge Tubes. Microwave excited electrodeless discharge tubes can be made for many elements. West and his co-workers have prepared them for at least fifty[24] and have published analytical data for a number. These data are summarized in Table 21. Where comparisons can be made, it is clear that this very intense source of radiation leads to an appreciable improvement in the detection limits.

TABLE 21. Fluorescence detection limits quoted by West and coworkers using electrodeless discharge tubes as primary sources

Element	Line nm	Detection limit p.p.m.	Flame gases	EDT power W	Ref.
Al	396.2	0.1	N_2O acet. (Ar-sep.)	40	102
As	235.0	0.15	Nitrogen hydrogen (diff.)	30	109
As	193.7	0.25	Nitrogen hydrogen (diff.)	30	109
As	193.7	1.0	Air acetylene	30	109
Be	234.9	0.01	N_2O acet. (slightly rich)	50	170
Bi	302.5	0.05	Nitrogen hydrogen (diff.)	12 (Iodine 206.2 nm)	108
Co	240.7	0.005	Air prop., air hydrogen	60	142
Co	240.7	0.01	Air acetylene	60	142
Fe	248.3	0.009	Air acetylene	60	127
Ge	265.1	1.1	N_2O acet. (N_2-sep.)	55	103
Ge	265.1	0.1	N_2O acet. (Ar-sep.)	55	103
Hg	253.7	0.08	Air hydrogen	20	76
Mo	313.3	0.05	N_2O acet. (Ar-sep.)	90	102
Sb	217.6	0.05	Air propane	30	107
Se	196.1	0.15	Air propane	30	106
Si	251.6	0.55	N_2O acet. (rich, Ar-sep.)	100	101
Sn	303.4	0.1	Oxygen argon hydrogen	40	75
Te	214.3	0.05	Air propane	20	106
Tl	377.6	0.12	Air hydrogen	50	76
V	318.4	0.07	N_2O acet. (Ar-sep.)	25	102
Zn	213.9	0.0001	Oxyhydrogen (direct injection)	—	96
Zn	213.9	0.002	Air acetylene	—	96

Non-flame atomization

The relative improvements in sensitivity which are a feature of the 'total sample' atomization methods described on p. 60 et seq. can also be attained in atomic fluorescence. The use of non-flame atom reservoirs has been discussed at some length by Winefordner.[450] Though there is little evidence so far of their application outside the research laboratory, there is little doubt that these enable a very great degree of sensitivity to be achieved.

So far, the Massmann furnace[254] and West's carbon filament[431] have been employed for this purpose. The accessibility of the atom cloud produced by the latter device would seem to make it ideal for fluorescence measurements, and detection limits of about three orders of magnitude better than for flame fluorescence have been reported. Actual limits of detection quoted[37] were magnesium and silver 10^{-12} g, lead and bismuth 10^{-11} g, thallium and gallium 5×10^{-11} g, and zinc 2×10^{-14} g. This last result then represented one of the best sensitivities ever achieved with an optical spectroscopic technique. With the filament unenclosed but shielded with argon Alder and West[30] found a detection limit of 1.5×10^{-13} g for cadmium.

General

At the present time, detection limits in fluorescence are superior to those in absorbance and flame emission for just a few elements—particularly the metalloids and the more volatile metals. This is because these elements make better intense sources such as electrodeless discharge tubes, and because they are atomized in the lower temperature, more transparent flames. These elements are not the most successful in absorption, and to this extent the two techniques are complementary.

Improvements in the fluorescence performance of the other elements are likely to result from more intense primary sources (or continua if a fluorescence instrument is to be freely versatile). But it seems unlikely that better sensitivities will be achieved with flame atomization, so analytical reproducibility and sensitivity may well have to be pursued in the non-flame atom sources.

Appendix 1

Sensitivities and other details for individual elements

The following information is tabulated by element: atomic number and atomic weight;

D_0, the dissociation energy of the oxide in eV*;

E_i, the ionization energy in eV*;

f, the oscillator strength*;

wavelengths of the strongest and other resonance lines suitable for use in atomic absorption analysis (D indicates a doublet);

gas system and type of flame (lean, stoichiometric, rich, etc. see Table 2, page 20) usually found to give the best sensitivity;

the approximate sensitivity value and, where possible, a typical detection limit in parts per million ($\mu g\ ml^{-1}$) in aqueous solution. Unless the source of the figure is given, the values quoted were obtained in the author's laboratory with a single beam instrument, the Unicam SP90. Using a Unicam SP1900, a double beam instrument of advanced design, improvements in the sensitivity of most elements by a factor of 2, and in detection limit by a factor of between 2 and 10, have been experienced;

major sources of interference;

useful extraction systems;

compounds which may be employed in making up non-aqueous standards. (The preparation of aqueous stocks and standards is detailed in Table 5, page 82.)

* Values collected by Rubeška and Moldan.[4]

Ag

SILVER

Atomic number 47 Atomic weight 107.868 D_0 1.4 E_i 7.57

λ	f	Flame	Approx. sens.	Det. limit
328.1	0.51	Air prop. lean	0.05	0.005
328.1	0.51	Air acet. lean	0.1	0.006
338.3	0.25	Air acet. lean	0.2	

Interferences

Most common elements do not interfere in air propane or air acetylene flames. Even finely suspended silver chloride gives the same response. Aluminium may cause slight depression and sulphuric and phosphoric acids cause depression through viscosity effects.

Extraction

(i) Di-*n*-butylammonium salicylate/MIBK[427],
(ii) Ammonium pyrrolidine dithiocarbamate/MIBK.

Compounds for making non-aqueous standards

Silver cyclohexane butyrate.
Silver 2-ethylhexanoate.

Al

ALUMINIUM

Atomic number 13 Atomic weight 26.98 D_0 5.0 E_i 5.98

λ	f	Flame	Approx. sens.	Det. limit
309.3(D)	0.23	N_2O acet. rich	0.8	0.2
308.2(D)	0.22	N_2O acet. rich	3	1
394.4	0.15	N_2O acet. rich	5	5

Interferences

Ionization: lanthanum or potassium may be used as an ionization buffer. Slight depression by calcium, silicon, and by perchloric and hydrochloric acids. Enhancement effects by titanium and iron and by fluoroboric and acetic acids.

Extraction

(i) Cupferron/MIBK.
(ii) 8-Quinolinol/chloroform.

Compounds for making non-aqueous standards

Aluminium cyclohexane butyrate.
Aluminium 2-ethylhexanoate.

As ARSENIC

Atomic number 33 Atomic weight 74.9216 D_0 4.9 E_i 9.81

λ	f	Flame	Approx. sens.	Det. limit
193.7	0.095	Air acet. just lum.	1.3	0.4
193.7	0.095	Argon hyd.*	0.6	0.3
197.2	0.07	Air acet. just lum.	3.0	0.5
197.2	0.07	Argon hyd.*	1.2	0.3
189.0	—	Argon hyd.*	—	—

***Flame**

Hydrogen 1500 cm^3/minute. Argon 6 1/minute.[187]

Interferences

Minor interferences with air acetylene flame. Major interferences from aluminium, barium, calcium, cobalt, iron, lithium, magnesium, nickel, potassium, silicon, sodium, and strontium in argon hydrogen flame. Scatter effects because of low wavelength of resonance lines.

Extraction

APDC/MIBK $\not>$ pH 4.

Au GOLD

Atomic number 79 Atomic weight 196.9665 E_i 9.22

λ	f	Flame	Approx. sens.	Det. limit
242.8	0.3	Air acet. lean	0.18	0.02
267.6	0.19	Air acet. lean	0.6	
274.8	—	Air acet. lean	250	

Interferences

Many interferences in air propane flame. Major interferences only from some other noble metals in air acetylene flame. Gold precipitates readily in reducing solutions, and this condition must be avoided in stock and working standards and samples. Cyanide complexes depress gold response, and must be destroyed by fuming with sulphuric–perchloric acid mixture.

Extraction

(i) APDC/MIBK.
(ii) β-Dimethylamino benzal rhodanine/amyl acetate (from aqua regia).

B BORON

Atomic number 5 Atomic weight 10.81 D_0 7.95δ E_i 8.29

λ	f	Flame	Approx. sens.
249.8	0.33	N_2O acet. rich	35
249.7	0.32	N_2O acet. rich	200

Interferences

No interference from common metals, in particular from 10 000 p.p.m. of iron, cobalt or nickel.

Extraction

2-Ethyl hexane 1,3-diol/chloroform.

Compounds for making non-aqueous standards

Menthyl borate.

Ba BARIUM

Atomic number 56 Atomic weight 137.34 D_0 5.95 E_i 8.29

λ	f	Flame	Approx. sens.	Det. limit
553.6	1.4	Air acet rich	7.0	
553.6	1.4	N_2O acet. rich	0.4*	0.2
455.4		N_2O acet. rich	2	1

***Interferences**

Potassium may be used as ionization buffer for 553.6 nm in the hot flame. Absorption interference from CaOH bands in presence of calcium is reduced in the nitrous oxide acetylene flame. This interference is largely obviated by measuring the barium ionic line 455.4 nm, which is sensitive only at the temperature of this flame. In this case, the ionization buffer should *not* be used . Barium can be separated from calcium by co-precipitation with lead chromate. Stable compound interference from phosphate, silicate, etc. in air acetylene flame is reduced in presence of lanthanum salts and also by use of the nitrous oxide acetylene flame.

Compounds for making non-aqueous standards

Barium cyclohexane butyrate.

Be BERYLLIUM

Atomic number 4 Atomic weight 9.01218 D_0 4.6 E_i 9.32

λ	f	Flame	Approx. sens.
234.9	0.24	N_2O acet. rich	0.05

Interferences

Some acids, particularly nitric, sulphuric, fluoroboric and acetic, enhance absorption by beryllium. Aluminium may depress absorption, but not in presence of hydrofluoric acid.

Bi BISMUTH

Atomic number 56 Atomic weight 208.9806 D_0 4.0 E_i 7.29

λ	f	Flame	Approx. sens.
223.1	0.012	Air prop. lean	0.5
223.1	0.012	Air acet. lean	0.8
222.8	0.0025	Air acet. lean	1.5
306.8	0.25	Air acet. lean	2.2
227.7	—	Air acet. lean	10

Interferences

No major interferences reported.

Extraction

APDC/MIBK pH 1–10.

Ca CALCIUM

Atomic number 20 Atomic weight 40.08 D_0 5.0 E_i 6.11

λ	f	Flame	Approx. sens.	Det. limit
422.7	1.49	Air acet. stoic.	0.06	0.005
422.7	1.49	N_2O acet. lean	0.03	0.005
239.9	0.037	Air acet. stoic	20	

Interferences

Many interferences have been reported in cool flames. In air acetylene, interferences occur with elements which form stable oxysalts, e.g. aluminium, beryllium, phosphorus, silicon, titanium, vanadium and zirconium. It is readily shown[316] that the depression only operates when the interferent is present in the sample solution, not if nebulized separately into the same flame, and it must therefore be a result of stable compound formation. The extent of the depression depends on the X/Ca ratio (where X is one of the above interferents) and is usually serious only at values greater than equivalence. Unfortunately such interferences are also strongest immediately above the blue cones of the air acetylene flame, i.e. where sensitivity is also best. The presence of organic solvents or other matter affects the position of maximum absorbance. These interferences are much reduced and often removed in the presence of a releasing agent such as lanthanum or strontium. They are not experienced in the nitrous oxide acetylene flame, though if this hot flame is employed potassium should also be present as ionization buffer.

Lanthanum itself depresses calcium slightly in air acetylene flames but enhances it 100% in nitrous oxide acetylene (the releasing agent should be added to samples and standards). There is some enhancement from sodium and potassium (ionization buffer effect) but acids, notably hydrofluoric, nitric, perchloric, sulphuric and trichloracetic decrease the absorption. The effect of protein bonding in biochemical samples is overcome by the use of EDTA buffer.

Extraction

8-Quinolinol/MIBK pH 11.

Compounds for making non-aqueous standards

Calcium 2-ethylhexanoate.

Cd CADMIUM

Atomic number 48 Atomic weight 112.40 D_0 3.8 E_i 6.11

λ	f	Flame	Approx. sens.	Det. limit
228.8	1.2	Air prop. lean	0.01	0.002
228.8	1.2	Air acet. lean	0.015	0.002
326.1	0.0018	Air acet. lean	20	

Interferences

None reported from common metals except silicon.

Extraction

APDC/MIBK.

Compounds for making non-aqueous standards

Cadmium cyclohexane butyrate.
Cadmium dibutyldithiocarbamate.

Ce CERIUM

Atomic number 58 Atomic weight 140.12

λ	Flame	Approx. sens.*
520.0(D)	N_2O acet.	30
569.7	N_2O acet.	39

* Varian Techtron: hollow cathode lamp technical data.

Interferences

Little is known about this element in atomic absorption. Because of the complexity of the cerium emission spectrum the likelihood of stray light interference in a normal monochromator bandpass is high.

Co COBALT

Atomic number 27 Atomic weight 58.9332 E_i 7.86

λ	f	Flame	Approx. sens.	Det. limit
240.7	0.22	Air acet. lean	0.08	0.02
242.5	0.19	Air acet. lean	0.2	
252.1	0.19	Air acet. lean	0.5	
341.3		Air acet. lean	4.0	

Interferences

Few chemical interferences reported, though 10000 p.p.m. of iron in perchloric acid enhance 10 p.p.m. by 25%. High relative concentrations of other elements should be investigated.

Extraction

APDC/MIBK pH 1–10.

Compounds for making non-aqueous standards

Cobalt cyclohexane butyrate.

Cr CHROMIUM

Atomic number 24 Atomic weight 51.996 D_0 4.2 E_i 6.76

λ	f	Flame	Approx. sens.	Det. limit
357.9	0.34	Air acet. rich	0.05	0.008
357.9	0.34	N_2O acet. rich	0.10	
359.3	0.27	Air acet. rich	0.15	
360.5	0.19	Air acet. rich	0.2	
425.4	0.10	Air acet. rich	0.5	

Interferences

Unabsorbed light interferes in low resolution monochromators. Chromium in samples and standards should be in same oxidation state e.g., either oxidized to Cr VI or reduced to Cr III. Major interference by iron in air acetylene flame reduced by addition of ammonium chloride. Depression by phosphate is overcome by addition of calcium. Minor interferences only in N_2O acetylene flame.

Extraction

(i) Cr VI (cold in presence of HCl)/MIBK.[140]
(ii) Diphenyl carbazone/MIBK (pH 0.5, H_2SO_4).[121]
(iii) APDC/MIBK (pH 3–7).

Compounds for making non-aqueous standards

Tris (1-phenyl-1,3-butanediono)chromium III.

Cs CAESIUM

Atomic number 55 Atomic weight 132.9005 E_i 3.87

λ	f	Flame	Approx. sens.[4]
852.1	0.8	Air acet. lean	0.5
455.5	—	Air acet. lean	20

Interferences

Stray light due to low photomultiplier sensitivity at 852 nm can be reduced by using a filter with a sharp cut-off below 800 nm. Ionization interferences: because caesium has lowest ionization potential of any metal a considerable excess, e.g. 0.1–0.5% of potassium must be used as ionization buffer. Some mineral acids depress absorbance slightly.

Cu COPPER

Atomic number 29 Atomic weight 63.54 D_0 4.9 E_i 7.72

λ	f	Flame	Approx. sens.	Det. limit
324.8	0.74	Air acet. lean	0.04	0.002
327.4	0.38	Air acet. lean	0.1	
217.9	0.011	Air acet. lean	0.6	
222.6	0.004	Air acet. lean	2.0	
249.2	—	Air acet. lean	10	
244.2	—	Air acet. lean	40	

Interferences

None reported except with very large excess of some elements, e.g. 10000 p.p.m. of iron depress 10 p.p.m. of copper by 10% in air acetylene flame.

Extraction

APDC/MIBK.

Compounds for making non-aqueous standards

Bis (1-phenyl-1,3-butanediono)copper II.
Copper cyclohexane butyrate.

Dy DYSPROSIUM

Atomic number 66 Atomic weight 162.50 E_i 6.2

λ	f	Flame	Approx. sens.[4]
421.2	—	N_2O acet. rich	1.5
404.6	—	N_2O acet. rich	1.8
419.5	—	N_2O acet. rich	3.4

Interferences

Ionization in high temperature flame, reduced by alkali metals and by some oxyacids.[245]

Er

ERBIUM

Atomic number 68 Atomic weight 167.26 E_i 6.2

λ	f	Flame	Approx. sens.[4]	Det. limit
400.8	—	N_2O acet. rich	1.4	0.1
415.1	—	N_2O acet. rich	2.0	

Interferences

Ionization reduced by addition of a large excess of an alkali metal.[245]

Eu

EUROPIUM

Atomic number 63 Atomic weight 151.96 E_i 5.64

λ	f	Flame	Approx. sens.	Det. limit
459.4	0.22	Acet. N_2O rich	2.0	0.2
462.7	0.20	Acet. N_2O rich	3.0	
466.2	0.17	Acet. N_2O rich	4.0	

Interferences

Ionization reduced by addition of potassium salts.

Fe IRON

Atomic number 26 Atomic weight 55.847 D_0 4.0 E_i 7.87

λ	f	Flame	Approx. sens.	Det. limit
248.3	0.34	Air acet. lean – stoic.	0.08	0.02
252.3	0.30	Air acet. lean – stoic.	0.3	
271.9	0.15	Air acet. lean – stoic.	0.5	
302.1	0.08	Air acet. lean – stoic.	0.7	
296.7	0.06	Air acet. lean – stoic.	1.0	
372.0	0.04	Air acet. lean —stoic.	1.0	
386.0	0.034	Air acet. lean – stoic.	2.0	
344.1	0.055	Air acet. lean – stoic.	5.0	
382.4	—	Air acet. lean – stoic.	30	

Interferences
Silica:[301] depression eliminated by addition of calcium chloride. Slight depression also in concentrated perchloric acid.
Organic acids:[332] the effect of citric acid is overcome by addition of phosphoric acid.

Extraction
(i) APDC/MIBK (pH 1–10).
(ii) Iron as a major constituent can be removed from Fe III chloride solution at pH 1 by isobutyl acetate or MIBK.

Compounds for making non-aqueous standards
Tris (1-phenyl-1,3-butanediono)iron III.

Ga — GALLIUM

Atomic number 31 Atomic weight 69.72 D_0 2.5 E_i 6.00

λ	f	Flame	Approx. sens.[282]	Det. limit
287.4	0.32	Air acet. lean	2.3	0.1
294.4	0.29	Air acet. lean	2.4	
417.2	0.14	Air acet. lean	3.7	
403.3	0.13	Air acet. lean	6.2	

Interferences

None reported.

Extraction

APDC/MIBK (pH 3–8).

Gd — GADOLINIUM

Atomic number 64 Atomic weight 157.25 D_0 6.0 E_i 6.16

λ	f	Flame	Approx. sens.[36]
368.4	—	N_2O acet. rich	40
407.9	—	N_2O acet. rich	—

Interferences

Ionization interference reduced by addition of potassium salts as ionization buffer.

Ge — GERMANIUM

Atomic number 32 Atomic weight 72.59 D_0 6.5 E_i 7.88

λ	f	Flame	Approx. sens.
265.2(D)	0.84	Acet. N_2O rich	1.5
265.2(D)	0.84	Air acet. lum.	5
259.3	0.37	Air acet. lum.	12
271.0	0.43	Air acet. lum.	12
275.4	0.22	Air acet. lum.	10

Hf

HAFNIUM

Atomic number 72 Atomic weight 178.49 E_i 6.8

λ	f	Flame	Approx. sens.[4]
307.3	0.02	N_2O acet. rich	14
286.6	—	N_2O acet. rich	30
289.8	—	N_2O acet. rich	70
296.4	—	N_2O acet. rich	80
368.2	—	N_2O acet. rich	—
377.8	—	N_2O acet. rich	—

Interferences

Persistence of stable Hf–O bonds in flame prevented by presence of HF. Iron also enhances slightly.

Hg

MERCURY

Atomic number 80 Atomic weight 200.59 E_i 10.43

λ	f	Flame	Approx. sens.	Det. limit
184.9*	—	—	—	—
253.7	0.03	Air propane lean	5	—
253.7	0.03	Air acet. lean	8	—
253.7	0.03	Monatomic vapour (no flame)	—	0.5 ng (see below)

* Just out of reach of most monochromators. Flame noise and atmospheric absorption are excessive but noise is said to be much reduced in an argon separated nitrous oxide acetylene flame.

Interferences

Many chemical interferences in both flames as these are of comparatively low temperature.

Extraction

APDC/MIBK, pH 0–10.

Cold vapour method for mercury

The poor sensitivity of mercury in the flame at 253.7 nm is overcome by generating monatomic mercury vapour chemically in the cold.[166] Alternatively mercury can be amalgamated electrolytically on to copper[73] or gold[404] and released as vapour by heating. Whichever way the mercury vapour is formed, it remains in the monatomic state at room temperature, and the absorption cell is a standard ultraviolet spectrometric gas cell.

In the simplest variation of the method the sample is dissolved or digested under oxidizing conditions chosen to prevent volatilization of the mercury as described on p. 174.

Apparatus and Method.[74] The apparatus depicted in Fig. 40 (p. 175) is set up. Air is circulated with a small diaphragm pump (Charles Austin Capex Mark II fitted with polypropylene inlet and outlet pipes) at about 1 litre per minute through, in the following order:

(i) the reaction vessel. This should have a volume convenient for the sample preparation procedures required (a 150 ml conical flask was suggested on p. 174) and should be fitted with a B24 ground glass socket to take a Drechsel bottle head;

(ii) a cotton wool and porosity G2 sinter spray trap;

(iii) the absorption cell, which should have the longest pathlength and minimum volume possible, consistent with transmission of sufficient energy to the detector. A length of 15 cm and internal diameter of 0.75 cm is suitable for most atomic absorption spectrometers. Alternatively a standard 10 cm ultraviolet gas-cell can be used, but this would cause the sensitivity to be worse, by a factor of about 2, than that otherwise possible. The burner itself acts as a convenient support for the cell, which can be fixed either with adhesive tape or with a specially made bracket;

(iv) and thence back to the pump.

The atomic absorption spectrometer is set up to make absorbance measurements on the mercury line 253.7 nm and the circulating pump is switched on.

The prepared sample solution, containing 0–300 ng of mercury as mercuric sulphate, is placed in the reaction vessel and the reductant, 2 ml of 10% stannous chloride solution, is added. The Drechsel bottle head is immediately inserted. The pump should then be switched on, when the mercury is swept in monatomic vapour form through the cell, The absorbance rises to a plateau, reaching a constant value in about one minute. When this constant value has been obtained and recorded, the reaction vessel is removed from the system and the mercury vapour is pumped out to the extraction hood. When the absorbance reaches its initial value, the next sample or standard can be reduced and inserted.

Sensitivity and Accuracy. Under the conditions described, an absorbance of unity should be obtained with about 1 μg of mercury in the system. As flame noise is completely absent, the measurements should be shot-noise limited and thus high scale expansion factors should be possible. The detection limit using this system should be less than 1 ng.

The repeatability of the measurements depends directly on the total volume of air circulating in the system. For a given analytical problem it is thus desirable to use flasks all of the same volume for the preparation of the samples and standards, and to ensure that the final volume of reaction mixture in the flask is also the same. A single reaction vessel cannot conveniently be used as not only may some mercury be lost on transfer, but there is a great danger of cross-contamination between samples.

An equally important consideration is that the mercury in the vapour phase is in equilibrium with the mercury in the liquid and the greater the volume of liquid for a given amount of mercury, the less mercury there will be in the vapour phase. If the liquid volume is variable in a system of constant total volume, these two sources of error tend, to some extent, to cancel each other. It is clear though, that for maximum sensitivity both the total volume of the system and the volume of liquid used should be as small as possible. For example, with a reaction vessel of volume 50 ml, an ultimate detection limit of about 0.1 ng is possible.

Interferences. Ions normally precipitating mercury from solution tend to depress or even completely suppress the evolution of mercury. Iodide, for example, causes such interference but is not likely to be present in samples which have been treated by the wet oxidation procedure. While this point should be remembered if water samples are being analysed without pretreatment, most metal concentrations up to 100 p.p.m. have not been found to interfere.

Ho HOLMIUM

Atomic number 67 Atomic weight 164.9303

λ	f	Flame	Approx. sens.	Det. limit
410.4	—	N_2O acet. rich	1.5	0.3
405.4	—	N_2O acet. rich	2.5	
416.3	—	N_2O acet. rich	3.5	
412.7	—	N_2O acet. rich	20	

Interferences

Ionization reduced by addition of potassium salts as ionization buffer.

In INDIUM

Atomic number 49 Atomic weight 114.82 D_0 1.1 E_i 5.78

λ	f	Flame	Approx. sens.[283]
303.9	0.36	Air acet. lean	0.9
325.6	0.37	Air acet. lean	0.9
410.5	0.14	Air acet. lean	2.6
271.0	—	Air acet. lean	—

Interferences

No serious interferences reported from most common metals and anions. Some depression from high magnesium and zinc.

Extraction

APDC/MIBK, pH 2–9.

Ir IRIDIUM

Atomic number 77 Atomic weight 192.22 E_i 9

λ	f	Flame	Approx. sens.	Det. limit
209.3	—	Air acet. stoic.	3	1.4
250.3	—	Air acet. stoic.	12	0.8
264.0	0.059	Air acet. stoic.	20	
254.4	—	Air acet. stoic.	60	

Extraction

APDC/MIBK. pH 1–14.

K POTASSIUM

Atomic number 19 Atomic weight 39.102 E_i 4.34

λ	f	Flame	Approx. sens.	Det. limit
766.5	0.69	Air propane lean	0.03	0.004
766.5	0.69	Air acet. lean	0.1	
769.9	0.34	Air acet. lean	0.2	
404.4	0.11	Air acet. lean	10.0	

Interferences

Ionization in air acetylene flame overcome by addition of sodium, lithium or caesium. There is also some depression by high mineral acid concentrations.

La LANTHANUM

Atomic number 57 Atomic weight 138.9055 D_0 7.0 E_i 5.61

λ	f	Flame	Approx. sens.
550.1	0.15	N_2O acet. rich	40
418.7	—	N_2O acet. rich.	50
357.4	0.12	N_2O acet. rich	100
392.7	0.18	N_2O acet. rich	200

Interferences

Cyanogen emission at 357.4 nm. Line at 550.1 is usually best.

Li LITHIUM

Atomic number 3 Atomic weight 6.941 E_i 5.39

λ	f	Flame	Approx. sens.
670.7	0.71	Air propane lean-stoic.	0.01
670.7	0.71	Air acet. lean-stoic.	0.03
323.3	0.026	Air acet. lean-stoic.	20

Interferences

Lithium is ionized in hot flames and may thus be affected by the presence of other alkali metals. An excess of potassium should be added as ionization buffer. Depressive effects caused by some strong acids are probably due to increased viscosity.

Compounds for making non-aqueous standards

Lithium cyclohexane butyrate.

Isotope determination

Lithium isotopes can be determined individually by atomic absorption, though one line of the doublet of Li^6 is superimposed on the other line of the doublet Li^7 (Manning and Slavin[247] and Svec and Anderson[382]).

Lu LUTECIUM

Atomic number 71 Atomic weight 174.97 D_0 4.3 E_i 6.15

λ	f	Flame	Approx. sens.
336.0	0.076	N_2O acet. rich	12[4]
328.2	0.086	N_2O acet. rich	100
308.1	0.096	N_2O acet. rich	—

Mg MAGNESIUM

Atomic number 12 Atomic weight 24.305 D_0 4.3 E_i 7.64

λ	f	Flame	Approx. sens.	Det. limit
285.2	1.2	Air acet. stoic.	0.005	0.0005
285.2	1.2	N_2O acet. lean-stoic.	0.005	
279.6 II	1.65	N_2O acet. lean-stoic.	0.2	
202.5	—	N_2O acet. lean-stoic.	25	

Interferences

Many interferences are reported in cool flames. Oxyacids of aluminium, phosphorus, silicon, titanium, etc., interfere in amounts greater than equivalence in air acetylene, and the effect is usually overcome by addition of releasing agents strontium or lanthanum. Most interferences are removed in the nitrous oxide acetylene flame, though that by titanium is persistent and may require lanthanum *and* the hotter flame.

Extraction

8-Hydroxyquinoline or 8-hydroxyquinaldine/MIBK in alkaline solution (e.g., pH 11 is often recommended).

Compounds for making non-aqueous standards

Magnesium cyclohexane butyrate.

Mn — MANGANESE

Atomic number 25 Atomic weight 54.9380 D_0 4 E_i 7.43

λ	f	Flame	Approx. sens.	Det. limit
279.5	0.58	Air acet. stoic.	0.025	0.002
403.1	—	Air acet. stoic.	0.5	
222.2	0.11	Air acet. stoic.	1.0	
321.7	—	Air acet. stoic.	100	

Interferences

Very few reported interferences with air acetylene flame. The effect of large excesses should be examined, e.g. 10000 p.p.m. of iron enhance 20 p.p.m. by 10% in an air acetylene flame. The depressive effect of silicon is overcome by addition of calcium.

Extraction

APDC/MIBK, pH 4–6 (pH 3 sometimes recommended).

Compounds for making non-aqueous standards

Manganous cyclohexane butyrate.

Mo MOLYBDENUM

Atomic number 42 Atomic weight 95.94 E_i 7.10

λ	f	Flame	Approx. sens.	Det. limit
313.3	0.2	Air acet. rich	0.5	0.1
313.3	0.2	N_2O acet. rich	0.2	
317.0	0.12	Air acet. rich	1.5	
379.8	0.13	Air acet. rich	2.0	
320.8	—	Air acet. rich	10.0	

Interferences

In an air acetylene flame, depression is caused by calcium, strontium, sulphate and iron, the last being serious. The extent depends on the observation height. These interferences are largely overcome by addition of ammonium chloride or aluminium chloride.[115, 279] There is no depression by iron in the nitrous oxide acetylene flame.

Extraction

APDC/MIBK, pH 2

Na SODIUM

Atomic number 11 Atomic weight 22.9898 E_i 5.14

λ	f	Flame	Approx. sens.	Det. limit
589.0(D)	0.76	Air propane lean	0.01	0.001
589.0(D)	0.76	Air acet. lean	0.02	
330.2(D)	0.055	Air acet. lean	5.0	

Interferences

High concentrations of calcium may interfere with Na 589.0 nm because of the noise caused by CaOH band emission. Ionization in hotter flames is overcome by the addition of potassium or other alkali metal. Few chemical interferences have been reported, though some strong mineral acids cause depression, particularly in the air propane flame.

Compounds for making non-aqueous standards

Sodium cyclohexane butyrate.
Sodium naphthasulphonate.
Sodium triphenylboron.

Nb NIOBIUM

Atomic number 41 Atomic weight 92.9064 D_0 4.0 E_i 6.88

λ	f	Flame	Approx. sens.
405.9	0.19	N_2O acet. rich	35
358.0	0.12	N_2O acet. rich	35
334.9	0.09	N_2O acet. rich	40

Interferences

Ionization. The above figures may be improved in presence of potassium, lanthanum or other element ionized in the nitrous oxide acetylene flame.

Considerable enhancements are experienced in hydrofluoric nitric acid solutions in the presence of iron. The effect reaches a plateau at 1 000–2 000 p.p.m. of iron for niobium concentrations 300–1 000 p.p.m. The presence of iron overcomes minor interferences by some common elements, e.g. nickel, cobalt, aluminium and chromium.

Extraction

APDC/MIBK. pH 2–4.

Nd NEODYMIUM

Atomic number 60 Atomic weight 144.24 E_i 5.45

λ	f	Flame	Approx. sens.[36]
463.4	0.08	N_2O acet. rich	35
471.9	—	N_2O acet. rich	73
492.5	0.09	N_2O acet. rich	—

Interferences

Ionization. The above sensitivities are improved in presence of potassium, lanthanum or other ionizable elements.

Ni NICKEL

Atomic number 28 Atomic weight 58.71 E_i 7.61

λ	f	Flame	Approx. sens.	Det. limit
232.0	0.095	Air acet. lean	0.1	0.02
234.6	0.051	Air acet. lean	0.5	
341.5	0.30	Air acet. lean	0.2	
346.2	0.16	Air acet. lean	1.0	
352.5	0.12	Air acet. lean	2.5	
339.1	—	Air acet. lean	5.0	
247.7		Air acet. lean	50	

Interferences

The line 232.0 nm requires a spectral slit width of about 0.2 nm, and if this is not available calibration graphs may be very curved. 341.5 is less interfered with in this way.

There are relatively few chemical interferences with nickel in the air acetylene flame particularly if the flame is lean but high concentrations should be examined. e.g. 10000 p.p.m. of iron enhance 10 p.p.m. of nickel by 10%. This effect is lower in the nitrous oxide acetylene flame.

Extraction

APDC/MIBK, pH 1–10 (pH 2–4 usually recommended).

Compounds for making non-aqueous standards

Nickel cyclohexane butyrate.

Os OSMIUM

Atomic number 76 Atomic weight 190.2 E_i 8.73

λ	f	Flame	Approx. sens.
290.9	—	N_2O acet. rich	1*
305.9	—	N_2O acet. rich	2
426.1	—	N_2O acet. rich	

* Sensitivities given by Willis in Slavin.[5]

Extraction

APDC/MIBK, pH 1–10

P PHOSPHORUS

Atomic number 15 Atomic weight 30.9738 E_i 10.48

λ	*f*	Flame	Approx. sens.
177.5	—	—	—

Extraction

Ammonium molybdate/iso-amyl alcohol—usually followed by indirect determination as molybdenum (See Chapter 5, p. 178.)

Compounds for making non-aqueous standards

Triphenyl phosphate.
For indirect methods of determining phosphorus see p. 177 *et seq.*

Pb LEAD

Atomic number 82 Atomic weight 207.2 D_0 4.1 E_i 7.42

λ	*f*	Flame	Approx. sens.	Det. limit
217.0	0.39	Air acet. lean	0.12	0.02
283.3	0.21	Air acet. lean	0.2	0.03
261.4	—	Air acet. lean	6.0	—

Interferences

Few interferences reported in air acetylene flame, though large excesses should be examined, e.g., 10 000 p.p.m. of iron enhance 5 p.p.m. of lead by 35%. Interferences previously reported in air propane flame in presence of elements which form refractory oxides are probably due to physical entrainment.

Extraction

APDC/MIBK, pH 0–8.

Compounds for making non-aqueous standards

Lead cyclohexane butyrate.

Pd PALLADIUM

Atomic number 46 Atomic weight 106.4 E_i 8.33

λ	f	Flame	Approx. sens.
247.6	0.1	Air acet. lean	0.5
244.8	0.074	Air acet. lean	1.0
276.3	0.071	Air acet. lean	3

Interferences
A few interferences, among them depression by strong mineral acids, have been reported for the air acetylene flame. Most metals (noble metals included) do not interfere.

Extraction
APDC/MIBK, pH 1–10

Pr PRASEODYMIUM

Atomic number 59 Atomic weight 140.9077 E_i 5.57

λ	f	Flame	Approx. sens.[245]
495.1	—	N_2O acet. rich	13
491.4	—	N_2O acet. rich	19
513.3	—	N_2O acet. rich	23
504.5	—	N_2O acet. rich	42

Pt PLATINUM

Atomic number 78 Atomic weight 195.09 E_i 9.0

λ	f	Flame	Approx. sens.
265.9	0.12	Air acet. lean-stoic.	2.5
306.5	—	Air acet. lean-stoic.	5.0
217.5	—	Air acet. lean-stoic.	10
299.8	—	Air acet. lean-stoic.	15

Interferences
Other noble metals give a complex pattern of interferences, which are lessened in the presence of 2% copper. Depression by strong mineral acids and some

common elements (e.g. magnesium) is also reported. There is enhancement in the presence of sodium sulphate (compare with rhodium).

Extraction

Extraction as $[PtCl_2 (SnCl_3)_2]^{2-}$ from hydrochloric acid solution into ethyl acetate.[297]

APDC/MIBK. pH 1–10.

Rb **RUBIDIUM**

Atomic number 37 Atomic weight 85.467 E_i 4.18

λ	f	Flame	Approx. sens.
780.0	0.80	Air acet. lean	0.5
794.8	0.40	Air acet. lean	1.0
420.2	—	Air acet. lean	10

Interferences

Ionization in air acetylene flame, reduced by addition of potassium or caesium, and reduced in the fuel rich flame. Depressive effect of some mineral acids can be minimized by choice of observation height, and compensated by matching of standards.

Re **RHENIUM**

Atomic number 75 Atomic weight 186.2 E_i 7.87

λ	f	Flame	Approx. sens.[36]
346.0	0.2	N_2O acet. rich	15
346.4	0.13	N_2O acet. rich	20
345.2	0.06	N_2O acet. rich	33

Interferences

Depressed in presence of some common elements particularly aluminium, manganese and calcium.

Extraction

APDC/MIBK, pH 1–10.

Rh

RHODIUM

Atomic number 45 Atomic weight 102.9055 E_i 7.45

λ	f	Flame	Approx. sens.	Det. limit
343.5	0.073	Air acet. lean	0.15	0.01
343.5	0.073	N_2O acet. lean	0.5	
339.6	—	Air acet. lean	0.5	
350.2	—	Air acet. lean	2.0	
328.1	—	Air acet. lean	15	

Interferences

Few metals interfere in the air acetylene flame, but anions, particularly those of mineral acids, have complex interfering effects which are overcome in presence of 1% of sodium sulphate.[190] Low concentrations of sulphate and nitrate both depress slightly. Interferences are reduced in the nitrous oxide acetylene flame, and calibration linearity is improved, though sensitivity is somewhat poorer.

Extraction

APDC/MIBK, pH 1–14.

Ru

RUTHENIUM

Atomic number 44 Atomic weight 101.07 E_i 7.34

λ	f	Flame	Approx. sens.[4]
349.9	0.10	Air acet. rich	2.0
372.8	0.087	Air acet. rich	—
392.6	—	Air acet. rich	—

Extraction

APDC/MIBK, pH 1–10.

Sb

ANTIMONY

Atomic number 51 Atomic weight 121.75 D_o 3.2 E_i 8.64

λ	f	Flame	Approx. sens.
206.8	0.1	Air acet. stoic.	0.5
217.6	0.045	Air acet. stoic.	1.1
231.2	0.03	Air acet. stoic.	1.9

Interferences

Most common elements do not interfere, though mineral acids may depress absorption. Standards should be matched for major ions.

Extraction

APDC/MIBK, pH 2–5.

Sc SCANDIUM

Atomic number 21 Atomic weight 44.9559 D_0 6.0 E_i 6.54

λ	f	Flame	Approx. sens.[36]
390.7	0.67	N_2O acet. stoic.	1.1
402.0	0.60	N_2O acet. stoic.	1.7
327.0	0.37	N_2O acet. stoic.	2.8

Se SELENIUM

Atomic number 34 Atomic weight 78.96 D_0 3.5 E_i 9.75

λ	f	Flame	Approx. sens.	Det. limit
196.1	0.12	Air acet. just luminous	0.8	0.2
196.1	0.12	Argon H_2	0.4	0.1
204.0	0.26	Argon H_2	—	—
206.3	0.30	Argon H_2	—	—

Interferences

Considerable freedom from interference in both air acetylene and argon hydrogen flames (Johns[187]).

Extraction

APDC/MIBK, pH 3–6.

Si SILICON

Atomic number 14 Atomic weight 28.086 E_i 8.15

λ	f	Flame	Approx. sens.	Det. limit
251.6	0.26	N_2O acet. rich	2.0	0.5
250.7	0.2	N_2O acet. rich	10	
251.4	0.54	N_2O acet. rich	10	
288.2	—	N_2O acet. rich	50	

Interferences

Ionization is decreased in the presence of a number of elements (including aluminium, calcium, sodium, vanadium etc.) and lanthanum, excess of which is added as an ionization buffer to maximize absorbance.

Extraction

As silicomolybdate into *n*-amyl alcohol.

Compounds for making non-aqueous standards

Silicone fluids.

Sm SAMARIUM

Atomic number 62 Atomic weight 150.4 E_i 5.6

λ	f	Flame	Approx. sens.[245]
429.7	—	N_2O acet. rich	20
520.1	—	N_2O acet. rich	30
528.3	—	N_2O acet. rich	50

Interferences

Ionization interferences. The above figures may be improved in presence of potassium.

Sn — TIN

Atomic number 50 Atomic weight 47.90 D_0 5.7 E_i 7.34

λ	f	Flame	Approx. sens.	Det. limit
224.6	0.41	N_2O acet. rich	0.8	0.2
286.3	0.23	N_2O acet. rich	2.5	0.5
286.3	0.23	Air acet. rich	10	
270.6	—	Air acet. rich	25	
303.4	—	Air acet. rich	50	

Flame

Air acetylene gives lower sensitivity and more interferences than nitrous oxide acetylene. An air hydrogen flame has been recommended, giving similar sensitivity to nitrous oxide acetylene but also with greater interferences.

Interferences

There are many contradictory statements about interferences affecting tin, but agreement that most are overcome in the nitrous oxide acetylene flame. High concentrations of concomitant elements should be examined, e.g., 10000 p.p.m. of iron enhance 20 p.p.m. of tin by 8%.

Extraction

APDC/MIBK, pH 4–6.

Sr — STRONTIUM

Atomic number 38 Atomic weight 87.62 D_0 4.85 E_i 5.69

λ	f	Flame	Approx. sens.	Det. limit
460.7	1.54	N_2O acet. stoic	0.04	0.006
460.7	1.54	Air acet. stoic	0.2	
407.8	0.76	Air acet. stoic	3.0	

Interferences

Similar to those on calcium. The effects of aluminium, phosphorus, silicon etc., in the air acetylene flame, are usually overcome in the presence of calcium or lanthanum. High concentrations of many anions and cations depress the response of strontium. These interferences are much reduced in the nitrous oxide acetylene flame, but an ionization buffer should be added to minimize the effects of ionization which is said to be 84% at the temperature of this flame.[36]

Compounds for making non-aqueous standards

Strontium cyclohexane butyrate.

Ta — TANTALUM

Atomic number 73 Atomic weight 180.9479 E_i 7.88

λ	f	Flame	Approx. sens.[36]
271.5	0.055	N_2O acet. rich	30
277.6	—	N_2O acet. rich	58
275.8	—	N_2O acet. rich	—

Tb — TERBIUM

Atomic number 65 Atomic weight 158.9254

λ	f	Flame	Approx. sens.[245]
432.7	—	N_2O acet. rich	8
431.9	—	N_2O acet. rich	9
433.9	—	N_2O acet. rich	—

Interferences

Ionization in nitrous oxide acetylene flame overcome by addition of potassium salts.

Te — TELLURIUM

Atomic number 52 Atomic weight 127.60 D_0 2.7 E_i 9.01

λ	f	Flame	Approx. sens.
214.3	0.08	Air acet. lean	0.5
225.9	—	Air acet. lean	—

Interferences

Few interferences are reported. There is the possibility of spectral interference from copper in the presence of this element when the lamp contains a cathode cup made of copper. Higher concomitant concentrations should be examined, and standards matched as necessary.

Extraction

APDC/MIBK. pH 3–5.

Th THORIUM

Atomic number 90 Atomic weight 232.0381 D_0 8.6 E_i 6.2

λ	f	Flame
371.9*	—	N_2O acet.
380.3	—	N_2O acet.
330.4	—	N_2O acet.

* Lines recommended in Varian Techtron hollow cathode lamp technical data. No sensitivities have been quoted.

Extraction

APDC/MIBK, pH 4–6.

Ti TITANIUM

Atomic number 22 Atomic weight 47.90 D_0 6.9 E_i 6.82

λ	f	Flame	Approx. sens.	Det. limit
364.3	0.25	N_2O acet. rich	1.5	0.4
365.4	0.22	N_2O acet. rich		
337.2	0.20	N_2O acet. rich		

Interferences

Titanium response can be enhanced in the presence of hydrofluoric acid, probably indicating the prevention of the formation of Ti—O bonds. An enhancement by iron in presence of hydrofluoric acid overcomes interference by a number of common elements.[167] In the absence of hydrofluoric acid but in perchloric acid solution. 10000 p.p.m. of iron have no effect on titanium absorption.

Compounds for making non-aqueous standards

Diethanolamine titanate.

Tl THALLIUM

Atomic number 81 Atomic weight 204.37 E_i 6.11

λ	f	Flame	Approx. sens.
276.8	0.27	Air acet. stoic.	0.2
377.6	0.13	Air acet. stoic.	—
238.0	0.07	Air acet. stoic.	—

Interferences

Small enhancements from many common elements equalized by addition of alkali metals to samples and standards.

Extraction

APDC/MIBK. pH 2–12.

Tm THULIUM

Atomic number 69 Atomic weight 168.9342 E_i 6.2

λ	f	Flame	Approx. sens.[4]
409.4	0.16	N_2O acet.	3.0
410.6	0.15	N_2O acet.	—
418.8	0.12	N_2O acet.	—
420.4	0.09	N_2O acet.	—

U URANIUM

Atomic number 92 Atomic weight 238.029 E_i 6.2

λ	f	Flame	Approx. sens.[36]
358.4	—	N_2O acet.	120
351.5	—	N_2O acet.	300
356.7	—	N_2O acet.	—

Interferences

Ionization in N_2O acetylene flame. Addition of ionization buffer would probably improve figures quoted.

Extraction

(i) With tributyl phosphate in carbon tetrachloride.
(ii) APDC/MIBK. pH 3–4.

V VANADIUM

Atomic number 23 Atomic weight 50.941 D_0 5.5 E_i 6.74

λ	f	Flame	Approx. sens.
318.39 (D)	0.66	N_2O acet. stoic.	1.0
318.34 (D)	0.5		
306.6	—	N_2O acet. stoic.	—
385.6	0.98	N_2O acet. stoic.	—
437.9	0.20	N_2O acet. stoic.	—

Interferences

Enhancement in presence of aluminium, titanium and a few common elements e.g., 10000 p.p.m. of iron enhance 100 p.p.m. of vanadium by 20%. Standards should thus be matched for major ions.

Extraction

(i) Cupferron MIBK.[341]
(ii) APDC/MIBK pH 4–6.

Compounds for making non-aqueous standards

Bis (1-phenyl-1,3-butanediono)oxovanadium IV.

W TUNGSTEN

Atomic number 74 Atomic weight 183.85 E_i 7.98

λ	f	Flame	Approx. sens.[36]
255.1	0.8	N_2O acet. rich	5
294.7	0.98	N_2O acet. rich	18
400.9	—	N_2O acet. rich	25

Interferences

Depression by major amounts of iron in perchloric acid not experienced in phosphoric–sulphuric–perchloric acid mixture.

Extraction

APDC/MIBK, pH 1–3.

Y YTTRIUM

Atomic number 39 Atomic weight 88.9059 D_0 7 E_i 6.51

λ	f	Flame	Approx. sens.	Det. limit
410.2	0.21	N_2O acet. slightly rich	3.0	1.0
412.8	0.18	N_2O acet. slightly rich	5.4	—
407.7	0.27	N_2O acet. slightly rich	5.7	—
414.3	0.20	N_2O acet. slightly rich	11	—

Interferences

Depression by a number of mineral acids (except nitric acid). Ionization minimized in presence of potassium.

Yb YTTERBIUM

Atomic number 70 Atomic weight 173.04 E_i 6.22

λ	f	Flame	Approx. sens.[36]
398.8	0.38	N_2O acet. stoic.	0.25
346.4	0.13	N_2O acet. stoic.	0.8
246.5	0.24	N_2O acet. stoic.	1.6

Interferences

Ionization in nitrous oxide acetylene flame overcome by addition of ionization buffer.

Zn

ZINC

Atomic number 30 Atomic weight 65.37 D_0 4 E_i 9.39

λ	f	Flame	Approx. sens.	Det. limit
213.9	1.2	Air acet. stoic.	0.012	0.001
307.6	0.00017	Air acet. stoic.	150	

Interferences

No serious interferences known. Slight depression by silicon.

Extraction

APDC/MIBK, pH 1–10 (pH 2.5–5 usually recommended).

Compounds for making non-aqueous standards

Zinc cyclohexane butyrate.

Zr

ZIRCONIUM

Atomic number 40 Atomic weight 91.22 D_0 7.8 E_i 6.84

λ	f	Flame	Approx. sens.[36]
360.1	0.22	N_2O acet. rich	20
352.0	—	N_2O acet. rich	—
301.2	—	N_2O acet. rich	34

Interferences

Improved response in presence of iron in hydrofluoric or hydrochloric acid solution.

Extraction

Cupferron/1 : 1 benzene–isoamyl alcohol.[403]

Appendix 2

Manufacturers of Atomic Absorption and Related Equipment

Manufacturer & Principal Address	U.K. Distributor	Types of instrument
Bausch & Lomb Inc., Rochester, N.Y., U.S.A.	Applied Research Laboratories, Luton, Beds.	Single beam
Beckman Instruments Inc., Fullerton, Calif., U.S.A.	Beckman–R.I.I.C., Croydon, Surrey	Single and double beam
Evans Electroselenium Ltd., Halstead, Essex, England		Single beam
Heath Schlumberger Benton Harbour, Mich., U.S.A.	Heath (Gloucester) Ltd., Gloucester	Single and double beam
Instrumentation Laboratory Inc., Lexington, Mass., U.S.A.	Instrumentation Laboratory (U.K.) Ltd., Altrincham, Cheshire	Double beam, dual channel
Jarrell Ash, Waltham, Mass, U.S.A.	V. A. Howe & Co., London	Single and double beam, dual channel
Jobin-Yvon, Longjumeau, France		Single beam
National Instrument Labs. Inc., Rockville, Maryland, U.S.A.		Single beam
Optica S.p.A., Milan, Italy		Single beam
Perkin Elmer Corporation, Norwalk, Conn., U.S.A.	Perkin Elmer Ltd., Beaconsfield, Bucks.	Single and double beam
Philips Electronic Instruments, Mt. Vernon, N.Y., U.S.A.		Single beam, multi-element
Pye Unicam Ltd., Cambridge, England		Single and double beam
Rank Precision Industries Ltd., Margate, Kent, England		Single beam
Shandon Southern Ltd., Camberley, England		Single beam
Shimadzu Seisakusho Ltd., Kyoto, Japan		Single beam
Spectrametrics Inc., Burlington, Mass., U.S.A.		Single beam
Varian Techtron Pty., Springvale, Vic., Australia	Varian Associates Ltd., Walton-on-Thames	Single beam
Carl Zeiss, Oberkochen, W. Germany	Degenhardt & Co. Ltd., London	Single beam

General Bibliography

BOOKS

1. Elwell, W. T. and Gidley, J. A. F., *Atomic Absorption Spectrophotometry*, Pergamon, Oxford, 1st ed., 1961, 2nd ed., 1966.
2. Robinson, J. W., *Atomic Absorption Spectroscopy*, Arnold, London, 1966.
3. Angino, E. E. and Billings, G. K., *Atomic Absorption Spectrometry in Geology*, Elsevier, Amsterdam, 1967.
4. Rubeška, I. and Moldan, B., *Atomic Absorption Spectrophotometry*, SNTL, Prague, 1967, English Edition, Iliffe, London, 1969.
5. Slavin, W., *Atomic Absorption Spectroscopy*, Interscience, New York, 1968.
6. Ramírez-Muñoz, J., *Atomic Absorption Spectroscopy*, Elsevier, Amsterdam, 1968.
7. Dean, J. A. and Rains, T. C., Eds., *Flame Emission and Atomic Absorption Spectrometry*, Vol. 1, *Theory*, Dekker, New York, 1969.
8. Christian, G. D. and Feldman, J. J., *Atomic Absorption Spectroscopy: Applications in Agriculture, Biology and Medicine*, Wiley-Interscience, New York, 1970.
9. Reynolds, R. J., Aldous, K. and Thompson, K. C., *Atomic Absorption Spectroscopy*, Griffin, London, 1970.
10. Rousselet, F., *Spectrophotométrie par Absorption Atomique appliquée à la Biologie*, Sedes, Paris, 1966.
11. L'vov, B. V., *Atomic Absorption Spectrochemical Analysis* (Translated from the Russian by J. H. Dixon) Adam Hilger Ltd., London, 1970.
12. Price, W. J., Chapters on atomic absorption and fluorescence in *Spectroscopy*, Ed.: D. R. Browning, McGraw Hill, London, 1969.

REVIEWS AND ABSTRACTS

13. *Atomic Absorption Newsletter*, Perkin Elmer Corporation, Norwalk, Conn., U.S.A.
14. *Spectrovision*, Pye Unicam Ltd., Cambridge, U.K.
15. *Flame Notes*, Beckman Instruments Inc., California, U.S.A.
16. *Analytical Abstracts*, Society for Analytical Chemistry, London, published monthly.
17. *Chemical Abstracts*, American Chemical Society.
18. *Spectrochemical Abstracts*, van Someren, E. H. S., Lachman, F. and Birks, F. T., Adam Hilger, London, published annually.
19. *Atomic Absorption and Flame Emission Spectroscopy Abstracts*, Masek, P. R., Sutherland, I. and Grivell, S., Science and Technology Agency, London, published bi-monthly.
20. Slavin, W., 'Atomic Absorption Spectroscopy, A Critical Review', *Appl. Spectry.*, **20**, 281 (1966).
21. West, T. S., 'Atomic Analysis in Flames', *Endeavour*, 44 (1967).
22. Reynolds, R. J., 'Atomic Absorption Spectroscopy, its Principles, Applications and Future Development', *Lab. Equip. Dig.* **6**, 79 (1968).

23. Slavin, W. and Slavin, S., 'Recent Trends in Analytical Atomic Absorption Spectroscopy', *Appl. Spectry.* **23**, 421 (1969).
24. West, T. S., 'Atom Flame Spectroscopy in Trace Analysis, Part 1: Atomic Absorption Spectroscopy', *Minerals Sci. Enging.* **1**, 3 (1969); Part 2: 'Atomic Fluorescence Spectroscopy', *Minerals Sci. Enging.* **2**, 31 (1970).
25. Platt, P., in *Annual Reviews in Analytical Chemistry*, S.A.C. London, 1971.
26. *Anal. Chem.* **38** (5) 305R (1966); **40** (5) 232R (1968); **42** (5) 206R (1970).
27. *Annual Reports on Analytical Atomic Spectroscopy*, S.A.C. London, 1972 onwards.
28. Reiss, R., *Atomic Fluorescence Bibliography*, Aztec Inst. Inc., South Norwalk, Conn., U.S.A., 1969.

REFERENCES

29. Adams, P. B. and Passmore, W. O., *Anal. Chem.* **38**, 630 (1966).
30. Alder, J. F. and West, T. S., *Anal. Chim. Acta* **51**, 365 (1970).
31. Alkemade, C. Th. J., *Appl. Opt.* **7**, 1261 (1968).
32. Allan, J. E., *Analyst* **86**, 530 (1961).
33. Allan, J. E., *Spectrochim. Acta* **17**, 459 (1961).
34. Allen, S. E. and Parkinson, J. A., *Spectrovision* **22**, 2 (1969).
35. Amos, M. D. and Thomas, P. E., *Anal. Chim. Acta* **32**, 139 (1965).
36. Amos, M. D. and Willis, J. B., *Spectrochim. Acta* **22**, 1325 (1966).
37. Anderson, G. R., Mains, I. S. and West, T. S., *Anal. Chim. Acta* **51**, 355 (1970).
38. Andrew, T. R. and Nichols, P. N. R., *Analyst* **87**, 25 (1962).
39. Antic-Jovanovic, A., Bojovic, V. and Marinkovic, M., *Spectrochim. Acta* **25B**, 405 (1970).
40. Atsuya, I., *Sci. Rep. Res. Inst., Tohoku Univ. Ser.* A **19**, 67 (1967).
41. Backer, E. T., *Clin. Chim. Acta* **24**, 233 (1969).
42. Bader, H. and Brandenburger, H., *At. Absorption Newsletter* **7**, 1 (1968).
43. Baker, C. A. and Garton, F. W. J., *U.K. At. Energy Authority, Rep.* R3490, H. M. Stationery Office, London (1961).
44. Barnes, L. Jr., *Anal. Chem.* **38**, 1083 (1966).
45. Barras, R. C. and Helwig, J. D., *Proc. Amer. Petrol. Inst.* Section III **43**, 223 (1963).
46. Baudin, G., Normand, J. and Fijalkowski, J., *Spectrochim. Acta* **23B**, 587 (1968).
47. Bazhov, A. S., *Zh. Analit. Khim.* **23**, 1640 (1968).
48. Beamish, F. E., Lewis, C. L. and van Loon, J. C., *Talanta* **16**, 1 (1969).
49. Bedrosian, A. J. and Lerner, M. W., *Anal. Chem.* **40**, 1104 (1968).
50. Belcher, C. B. and Bray, H. M., *Anal. Chim. Acta* **26**, 322 (1962).
51. Belcher, C. B. and Kinson, K., *Anal. Chim. Acta* **30**, 483 (1964).
52. Bell, G. F., *At. Absorption Newsletter* **5**, 73 (1966).
53. Belt, C. B. Jr., *Anal. Chem.* **39**, 676 (1967).
54. Berge, D. G., Pflaum, R. T., Lehman, D. A. and Frank, C. W., *Anal. Letters* **1**, 613 (1968).
55. Berman, E., *At. Absorption Newsletter* **6**, 57 (1967).
56. Bernas, B., *Anal. Chem.* **40**, 1682 (1968).
57. Biechler, D. G., *Anal. Chem.* **37**, 1054 (1965).
58. Billings, G. K. and Ragland, P. C., *Can. Spectry.* **14**, 8 (1969).
59. Black, C. A., et al., Eds. *Methods of Soil Analysis* American Society of Agronomy (1965).
60. Blakemore, L. C., *New Zealand J. Agr. Res.* **11**, 515 (1968).
61. Blijenberg, B. G. and Leijnse, B., *Clin. Chim. Acta* **19**, 97 (1968).
62. Blomfield, J., McPherson, J. and George, C. R. P., *Brit. Med. J.* **2**, 141 (1969).
63. Boar, P. L. and Ingram, L. K., *Analyst* **95**, 124 (1970).
64. Boettner, E. A. and Grunder, F. I., in *Trace Inorganics in Water: Advances in Chemistry Series* No. 73, American Chemical Society, Washington D.C., 1968.
65. Bokowski, D. L., *Amer. Ind. Hyg. Assoc. J.* **29**, 474 (1968).

66. Boling, E. A., *Spectrochim. Acta* **22**, 425 (1966).
67. Bond, A. M. and O'Donnell, T. A., *Anal. Chem.* **40**, 560 (1968).
68. Bond, A. M. and Willis, J. B., *Anal. Chem.* **40**, 2087 (1968).
69. Bowman, J. A., *Anal. Chim. Acta* **37**, 465 (1967).
70. Bowman, J. A. and Willis, J. B., *Anal. Chem.* **39**, 1210 (1967).
71. Bradfield, E. G. and Spencer, D., *J. Sci. Food Agr.* **16**, 33 (1965).
72. Brandenburger, H., *Ann. Biol. Clin. (Paris)* **25**, 1053 (1967).
73. Brandenberger, von H. and Bader, H., *Helv. Chim. Acta* **50**, 1409 (1967).
74. Braun, R. and Husbands, A. P., *Spectrovision* **26**, 2 (1971).
75. Browner, R. F., Dagnall, R. M. and West, T. S., *Anal. Chim. Acta* **46**, 207 (1969).
76. Browner, R. F., Dagnall, R. M. and West, T. S., *Talanta* **16**, 75 (1969).
77. Bryce-Smith, D., *Chem. Brit.* **7**, 54 (1971).
78. Burrows, J. A., Heerdt, J. C. and Willis, J. B., *Anal. Chem.* **37**, 579 (1965).
79. Butler, F. E., *Amer. Ind. Hyg. Assoc. J.* **30**, 559 (1969).
80. Butler, L. R. P. and Fulton, A., *Appl. Opt.* **7**, 2131 (1968).
81. Butler, L. R. P. and Strasheim, A., *Spectrochim. Acta* **21**, 1207 (1965).
82. Capacho-Delgado, L. and Manning, D. C., *At. Absorption Newsletter* **5**, 1 (1966).
83. Capacho-Delgado, L. and Manning, D. C., *Spectrochim. Acta* **22**, 1505 (1966).
84. Chao, T. T., Fishman, M. J. and Ball, J. W., *Anal. Chim. Acta* **47**, 189 (1969).
85. Chau, Y. K. and Lum-Shue-Chan, K., *Anal. Chim. Acta* **48**, 205 (1969).
86. Chester, J. E., Dagnall, R. M. and Taylor, M. R. G., *Analyst* **95**, 702 (1970).
87. Christian, G. D., *Clin. Chem.* **11**, 459 (1965).
88. Christian, G. D. and Feldman, F. J., *Anal. Chim. Acta* **40**, 173 (1968).
89. Clarke, W. E. and Cooke, P. A., *BCIRA Rep.* No. 891 (1967).
90. Collinson, W. J. and Boltz, D. F., *Anal. Chem.* **40**, 1896 (1968).
91. Cooke, P. A. and Price, W. J., *Spectrovision* **16**, 7 (1966).
92. Cotton, D. H. and Jenkins, D. R., *Spectrochim. Acta* **25B**, 283 (1970).
93. Crawford, L. R. Jr and Greweling, T., *Appl. Spectry.* **22**, 793 (1968).
94. Cresser, M. S. and Keliher, P. N., *Amer. Lab.* p 8 (Aug. 1970).
95. Cresser, M. S. and Keliher, P. N., *Intern. Lab.* p. 17 (Jan./Feb. 1971).
96. Cresser, M. S. and West, T. S., *Anal. Chim. Acta* **50**, 517 (1970).
97. Cresser, M. S. and West, T. S., *Spectrochim. Acta* **25B**, 61 (1970).
98. Cunningham, A. F., *At. Absorption Newsletter* **8**, 70 (1969).
99. Curnow, D. H., Gutteridge, D. H. and Horgan, E. D., *At. Absorption Newsletter* **7**, 45 (1968).
100. Curry, A. S., Read, J. F. and Knott, A. R., *Analyst* **94**, 744 (1969).
101. Dagnall, R. M., Kirkbright, G. F., West, T. S. and Wood, R., *Anal. Chim. Acta* **47**, 407 (1969).
102. Dagnall, R. M., Kirkbright, G. F., West, T. S. and Wood, R., *Anal. Chem.* **42**, 1029 (1970).
103. Dagnall, R. M., Kirkbright, G. F., West, T. S. and Wood, R., *Analyst* **95**, 425 (1970).
104. Dagnall, R. M., Thompson, K. C. and West, T. S., *Anal. Chim. Acta* **36**, 269 (1966).
105. Dagnall, R. M., Thompson, K. C. and West, T. S., *Talanta* **14**, 551 (1967).
106. Dagnall, R. M., Thompson, K. C. and West, T. S., *Talanta* **14**, 557 (1967).
107. Dagnall, R. M., Thompson, K. C. and West, T. S., *Talanta* **14**, 1151 (1967).
108. Dagnall, R. M., Thompson, K. C. and West, T. S., *Talanta* **14**, 1467 (1967).
109. Dagnall, R. M., Thompson, K. C. and West, T. S., *Talanta* **15**, 677 (1968).
110. Dagnall, R. M. and West, T. S., *Appl. Opt.* **7**, 1287 (1968).
111. Dagnall, R. M., West, T. S. and Young, P., *Anal. Chem.* **38**, 358 (1966).
112. Dalton, E. F. and Melanoski, A. J., *J. Assoc. Offic. Anal. Chemists* **52**, 1035 (1969).
113. David, D. J., *Analyst* **89**, 747 (1964).
114. David, D. J., *Analyst* **87**, 576 (1962).
115. David, D. J., *Analyst* **86**, 730 (1961).
116. David, D. J., *Analyst* **85**, 495 (1960).
117. David, D. J., *Analyst* **84**, 536 (1959).
118. David, D. J., *Analyst* **83**, 655 (1958).

119. Dawson, J. B. and Heaton, F. W., *Spectrochemical Analysis of Clinical Materials* Thomas, Springfield, U.S.A. (1967).
119*a* Dawson, J. B. and Heaton, F. W., *Biochem. J.* **80**, 99 (1961).
120. Deily, J. R., *At. Absorption Newsletter* **5**, 119 (1966).
121. Delaughter, B., *At. Absorption Newsletter* **4**, 273 (1965).
122. Delves, H. T., *Biochem. J.* **112**, 34P (1969).
123. Delves, H. T., *Analyst* **95**, 431 (1970).
124. Demers, D. R. and Mitchell, D. C., *Technicon International Congress*, Nov. 2–4, 1970.
125. Duncan, G. and Herridge, R. J., *Talanta* **17**, 766 (1970).
126. Dyck, R., *At. Absorption Newsletter* **4**, 170 (1965).
127. Ebdon, L., Kirkbright, G. F. and West, T. S., *Anal. Chim. Acta* **47**. 563 (1969).
128. Eisen, J., *Z. Erzbergbau Metallheutten.* **16**. 579 (1963).
129. Elenbaas, W. and Riemans, J., *Philips Tech. Rev.* **11**, 299 (1950).
130. Elwell, W. T. and Gidley, J. A. F., *Anal. Chim. Acta* **24**, 71 (1961).
131. Elwell, W. T. and Scholes, I. R., *Analysis of Copper and its Alloys* Pergamon. Oxford. 1967.
132. Elwell, W. T. and Wood, D. F., *Analysis of the New Metals*, Pergamon, Oxford (1966).
133. Erinc, G. and Magee, R. J., *Anal. Chim. Acta* **31**, 197 (1964).
134. Farrar, B., *At. Absorption Newsletter* **4**, 325 (1965).
135. Farrelly, R. O. and Pybus, J., *Clin. Chem.* **15**, 566 (1969).
136. Feldman, F. J., *Research/Development* p. 22 (Oct. 1969).
137. Feldman, F. J., *Anal. Chem.* **42**, 719 (1970).
138. Feldman, F. J., Blasi, J. A. and Smith, S. B. Jr., *Anal. Chem.* **41**, 1095 (1969).
139. Feldman, F. J., Knoblock, E. C. and Purdy, W. C., *Anal. Chim. Acta* **38**. 489 (1967).
140. Feldman, F. J. and Purdy, W. C., *Anal. Chim. Acta* **33**, 273 (1965).
141. Fishman, M. J. and Midgett, M. E., in *Trace Inorganics in Water: Advances in Chemistry Series* No. 73, American Chemical Society, Washington, D.C. (1968).
142. Fleet, B., Liberty, K. V. and West, T. S., *Anal. Chim. Acta* **45**. 205 (1969).
143. Ford, M. A., *Photoelec. Spectrometry Group Bull.* **18**, 554 (1968).
144. Frey, S. W. Dewitt, W. G. and Bellomy, B. R., *Amer. Soc. Brewing Chemists Proc.* 172 (1966).
145. Frey, S. W., Dewitt, W. G. and Bellomy, B. R., *Amer. Soc. Brewing Chemists Proc.* 199 (1967).
146. Fuwa, K. and Vallee, B., *Anal. Chem.* **35**, 942 (1963).
147. Fuwa. K. and Vallee. B., *Anal. Chem.* **41**, 188 (1969).
148. de Galan, L. and Samaey, G. F., *Spectrochim. Acta.* **25B**, 245 (1970).
149. Gamot, E., Philibert, J. and Vialette, Y., *Colloq. Nationaux Centre Na. Rech. Sci.* Nancy, 4–6 decembre 1968, p. 287.
150. Ganivet, M. and Benhamou, A., *Anal. Chim. Acta* **47**, 81 (1969).
151. Gatehouse, B. M. and Walsh, A., *Spectrochim. Acta* **16**, 602 (1960).
152. Gaydon, A. G. and Wolfard, H. G., *Flames, their Structure, Radiation and Temperature*, Chapman and Hall, London, p. 304 (1960).
153. Gidley, J. A. F. and Jones, J. T., *Analyst* **85**, 249 (1960).
154. Gimblet, E. G., Marney, A. G. and Bonsnes, R. W., *Clin. Chem.* **13**. 204 (1967).
155. Ginzburg, V. L., Livshits, D. M. and Satarina, G. I., *Zh. Analit. Khim.* **19**, 1089 (1964).
156. Goecke, R., *Talanta* **15**, 871 (1968).
157. Goecke, R. F., *At. Absorption Newsletter* **8**, 106 (1969).
158, Grant, C. L., *Atomic Absorption Spectroscopy* A.S.T.M. Special Technical Publication 443, p. 37 (1969).
159. Hale, B. J., *J. Agr. Sci.* **37**, 236 (1947).
160. Halstead, E. H., Barber, S. A., Warncke, D. D. and Bole, J. B.. *Proc. Amer. Soil Sci. Soc.* **32**, 69 (1968).
161. Halstead, R. L., Finn, B. J. and MacLean, A. J., *Can. J. Soil.* **49**, 335 (1969).
162. Hansen, J. L., *Amer. J. Med. Technol.* **34**, 625 (1968).
163. Hardy, H. L., *J. Occupational. Med.* **1**, 220 (1959).

164. Harrison, W. W., Yuracheck, J. P. and Benson, C. A., *Clin. Chim. Acta* **23**, 83 (1969).
165. Hartley, F. R. and Inglis, A. S., *Analyst* **92**, 622 (1967).
166. Hatch, W. R. and Ott, W. L., *Anal. Chem.* **40**, 2085 (1968).
167. Headridge, J. B. and Hubbard, D. P., *Anal. Chim. Acta* **37**, 151 (1967).
168. Hickey, L. G., *Anal. Chim. Acta* **41**, 546 (1968).
169. Heffernan, B. J., Archibold, R. O. and Vickers, T. J., *Australasian Inst. Mining and Met. Proc.* **223**, 65 (1967).
170. Hingle, D. N., Kirkbright, G. F. and West, T. S., *Analyst* **93**, 522 (1968).
171. Hingle, D. N., Kirkbright, G. F. and West, T. S., *Talanta* **15**, 199 (1968).
172. Hirano, S., Tofuku, Y. and Fujii, T., *Japan Analyst* **18**, 574 (1969).
173. Hofer, A., *Z. Anal. Chem.* **249**, 115 (1970).
174. Holak, W., *Anal. Chem.* **41**, 1712 (1969).
175. Hoover, W. L., Reagor, J. C. and Garner, J. C., *J. Assoc. Offic. Anal. Chemists* **52**, 708 (1969).
176. Hornbrook, E. H. W., *Can. Mining J.* **90**, 107 (1969).
177. Hubbard, D. P. and Monks, H. H., *Anal. Chim. Acta* **47**, 197 (1969).
178. Hullin, R. P., Kapel, M. and Drinkall, J. A., *J. Food Technol.* **4**, 235 (1969).
179. Humphrey, J. R., *Anal. Chem.* **37**, 1604 (1965).
180. Hurford, T. R. and Boltz, D. F., *Anal. Chem.* **40**, 379 (1968).
181. Husler, J., *At. Absorption Newsletter* **9**, 31 (1970).
182. Hwang, J. Y. and Sandonato, L. M., *Anal. Chem.* **42**, 744 (1970).
183. Ishii, T., Musha, S. and Munemori, M., *Japan Analyst* **17**, 27 (1968).
184. Ito, J., *Bull. Chem. Soc. Japan* **35**, 225 (1962).
185. Jakubiec, R. J. and Boltz, D. F., *Anal. Letters* **1**, 347 (1968).
186. Jimenez Seco, J. L. and Coldo, A. G., *Rev. Met.* **4**, 621 (1968).
187. Johns, P., *Spectrovision* **24**, 6 (1970).
188. Johns, P., *Spectrovision* **26**, 15 (1971).
189. Johns, P. and Price, W. J., *Metallurgia*, **81**, 75 (1970).
190. Johns, P. and Price, W. J., *Pittsburgh Conference*, March 1970, Abstract 108.
191. Jolly, S. C. (Ed.) (1) *Official Standardised and Recommended methods of Analysis*, London 1963; (2) *Supplement to Official, Standardised and Recommended Methods of Analysis*, Society for Analytical Chemistry, London 1967.
192. Jordan, J., *At. Absorption Newsletter* **7**, 48 (1968).
193. Jursik, M. L., *At. Absorption Newsletter* **6**, 21 (1967).
194. Kahn, H. L., *At. Absorption Newsletter* **6**, 51 (1967).
195. Kahn, H. L., Peterson, G. E. and Schallis, J. E., *At. Absorption Newsletter* **7**, 35 (1968).
196. Kahn, H. L. and Schallis, J. E., *At. Absorption Newsletter* **7**, 5 (1968).
197. Kapetan, J. P., *Atomic Absorption Spectroscopy* A.S.T.M. Special Publication 443 1969, p. 78.
198. Kashiki, M., Yamazoe, S. and Oshima, S., *Anal. Chim. Acta* **53**, 95 (1971).
199. Kerber, J. D., *Appl. Spectry.* **20**, 212 (1966).
200. Kinnunen, J. and Lindsjö, O., *Chemist-Analyst* **56**, 25 and 76 (1967).
201. Kinson, K., Hodges, R. J. and Belcher, C. B., *Anal. Chim. Acta* **29**, 134 (1963).
202. Kinson, K. and Belcher, C. B., *Anal. Chim. Acta* **30**, 64 (1964).
203. Kinson, K. and Belcher, C. B., *Anal. Chim. Acta* **31**, 180 (1964).
204. Kirchhoff, G. and Bunsen, R., *Pogg. Ann.* **110**, 161 (1860); **113**, 337 (1861); *Phil. Mag.* **22**, 329 (1861).
205. Kirkbright, G. F., Peters, M. K. and West, T. S., *Talanta* **14**, 789 (1967).
206. Kirkbright, G. F., Peters, M. K. and West, T. S., *Analyst* **91**, 411 (1966).
207. Kirkbright, G. F., Semb, A. and West, T. S., *Talanta* **15**, 441 (1968).
208. Kirkbright, G. F., Smith, A. M. and West, T. S., *Analyst* **91**, 700 (1966).
209. Kirkbright, G. F., Smith, A. M. and West, T. S., *Analyst* **92**, 411 (1967).
210. Kirkbright, G. F., Smith, A. M. and West, T. S., *Analyst* **93**, 292 (1968).
211. Kirkbright, G. F., Smith, A. M., West, T. S. and Wood, R., *Analyst* **94**, 754 (1969).
212. Kirkbright, G. F. and West, T. S., *Appl. Opt.* **7**, 1305 (1968).

213. Knight, D. M. and Pyzyna, M. K., *At. Absorption Newsletter* **8**, 129 (1969).
214. Kohlenberger, D. W., *At. Absorption Newsletter* **8**, 108 (1969).
215. Koirtyohann, S. R. and Feldman, C., in Forette, J. E. and Lanterman E. (Eds.), *Developments in Applied Spectroscopy* Vol. 3, Plenum Press, New York, 1964.
216. Kometani, T. Y., *Plating*, 1251 (Nov. 1969).
217. König, P., Schmitz, K. H. and Thiemann, E., *Z. Anal. Chem.* **244**, 232 (1969).
218. Kopito, L. and Shwachman, H., *J. Lab. Clin. Med.*, **70**, 326 (1967).
219. Kumamaru, T., *Anal. Chim. Acta* **43**, 19 (1968).
220. Laflamme, Y., *At. Absorption Newsletter* **7**, 101 (1968).
221. Langmyhr, F. J. and Graff, P. R., *Anal. Chim. Acta* **21**, 334 (1959).
222. Langmyhr, F. J. and Paus, P. E., *Anal. Chim. Acta* **43**, 397 (1968).
223. Langmyhr, F. J. and Paus, P. E., *Anal. Chim. Acta* **43**, 506 (1968).
224. Langmyhr, F. J. and Paus, P. E., *Anal. Chim. Acta* **43**, 508 (1968).
225. Langmyhr, F. J. and Paus, P. E., *Anal. Chim. Acta* **44**, 445 (1969).
226. Langmyhr, F. J. and Paus, P. E., *Anal. Chim. Acta* **45**, 173 (1969).
227. Langmyhr, F. J. and Paus, P. E., *Anal. Chim. Acta* **45**, 176 (1969).
228. Langmyhr, F. J. and Paus, P. E., *Anal. Chim. Acta* **45**, 157 (1969).
229. Langmyhr, F. J. and Paus, P. E., *Anal. Chim. Acta* **50**, 515 (1970).
230. Langmyhr, F. J. and Sveen, S., *Anal. Chim. Acta* **32**, 1 (1965).
231. Law, S. L. and Green, T. E., *Anal. Chem.* **41**, 1008 (1969).
232. Leaton, J. R., *J. Assoc. Offic. Anal. Chemists* **53**, 237 (1970).
233. Lehmann, V., *Clin. Chim. Acta* **20**, 523 (1968).
234. Lehnert, G., Klavis, G., Schaller, K. H. and Haas, T., *Brit. J. Ind. Med.* **26**, 156 (1969).
235. Lehnert, G. and Schaller, K. H., *Med. Welt*, N.F. **18**, 1131 (1967).
236. Lewis, C. L., Ott, W. L. and Sine, N. M., *The Analysis of Nickel*, Pergamon, Oxford, 1966.
237. Lewis, L. L., *Atomic Absorption Spectroscopy*, A.S.T.M. Special Technical publication 443, 1969, p. 47.
238. Lindstedt, G., *Analyst* **95**, 264 (1970).
239. Lorber, A., Cohen, R. L., Chang, C. C. and Anderson, H. E., *Arthritis Rheumat.* **11**, 170 (1968).
240. L'vov, B. V., *Spectrochim. Acta* **17**, 761 (1961).
241. L'vov, B. V., *Spectrochim. Acta* **24B**, 53 (1969).
242. L'vov, B. V., *Atomic Absorption Spectroscopy* Plenary Lectures of International Conference, Sheffied, July 1969, Butterworths, London, 1970, p. 11.
243. Mahoney, J. P., Sargent, K., Greland, M. and Small, W., *Clin. Chem.* **15**, 312 (1969).
244. Malissa, H. and Schöffmann, E., *Mikrochim. Acta* (1) 187 (1955).
245. Manning, D. C., *At. Absorption Newsletter* **5**, 127 (1966).
246. Manning, D. C. and Fernandez, F., *At. Absorption Newsletter* **9**, 65 (1970).
247. Manning, D. C. and Slavin, W., *At. Absorption Newsletter* No. 8 (Nov. 1962).
248. Mansell, R. E., Emmel, H. W. and McLaughlin, E. L., *Appl. Spectry.* **20**, 231 (1966).
249. Marcec, M. V., Kinson, K. and Belcher, C. B., *Anal. Chim. Acta* **41**, 447 (1968).
250. Marks, J. Y. and Welcher, G. G., *Anal. Chem.* **42**, 1033 (1970).
251. Marshall, G. B. and West, T. S., *Talanta* **14**, 823 (1967).
252. Marshall, G. B. and West, T. S., *Analyst* **95**, 343 (1970).
253. Massmann, H., *Method. Phys. Anal.* **4**, 193 (1968).
254. Massmann, H., *Spectrochim. Acta* **23B**, 215 (1968).
255. Matousek, J. and Sychra, V. *Anal. Chem.* **41**, 518 (1969).
256. Matsuo, T., Shida, J. and Motoki, M. *Japan Analyst* **18**, 521 (1969).
257. Matousek, J. and Sychra, V. *Anal. Chim. Acta* **49**, 175 (1970).
258. Mavrodineanu, R. and Boiteux, H., *Flame Spectroscopy* Wiley, New York, 1965.
259. Mavrodineanu, R. and Hughes, R. C., *Appl. Opt.* **7**, 1281 (1968).
260. McCrackan, J. D., Vecchione, M. C. and Longo, S. L., *At. Absorption Newsletter* **8**, 102 (1969).
261. McIsaac, C. L., *Eng. Mining J.* **170**, 55 (1969).
262. McLean, A. J., Halstead, R. L. and Finn, B. J., *Can. J. Soil Sci.* **49**, 327 (1969).

263. MacPhee, W. S. G. and Ball, D. F., *J. Sci. Food Agr.* **18**, 376 (1967).
264. McPherson, G. L., *4th Australian Spectroscopy Conference, Canberra* Aug. 1962.
265. Means, E. A. and Ratcliffe, D., *At. Absorption Newsletter* **4**, 174 (1965).
266. Mee, J. M. L. and Hilton, H. W., *J. Agr. Food Chem.* **17**, 1398 (1969).
267. Melton, J. R., Hoover, W. L. and Howard, P. A., *J. Assoc. Offic. Anal. Chemists* **52**, 950 (1969).
268. Menis, O. and Rains, T. C., *Anal. Chem.* **41**, 952 (1969).
269. Meranger. J. C. and Somers, E., *Analyst* **93**, 799 (1968).
270. Meredith, M. K., Baldwin, S. and Andreasen, A. A., *J. Assoc. Offic. Anal. Chemists* **53**, 12 (1970).
271. Mitchell, A. C. G. and Zemansky, M. W., *Resonance Radiation and Excited Atoms* Cambridge University Press, London, 1934, reprinted 1961.
272. Mitchell, D. G., *Technicon International Congress*, Nov. 2–4 1970.
273. Mitchell, D. G. and Johansson, A., *Spectrochim. Acta* **25B**, 175 (1970); **26B**, 677 (1971).
274. Mizuno, T., Harada, A., Kudo, Y. and Hasegawa, N., *Japan Analyst* **19**, 251 (1970).
275. Montford, B. and Cribbs, S. C., *At. Absorption Newsletter* **8**, 77 (1969).
276. Montford, B. and Cribbs, S. C., *Talanta* **16**, 1079 (1969).
277. Moore, E. J., Milner, O. I. and Glass, J. R., *Microchem. J.* **10**, 148 (1966).
278. Morrison, G. H. and Freiser, H., *Solvent Extraction in Analytical Chemistry* Wiley, New York, 1957.
279. Mostyn, R. A. and Cunningham, A. F., *Anal. Chem.* **38**, 121 (1966).
280. Mostyn, R. A. and Cunningham, A. F., *At. Absorption Newsletter* **6**, 86 (1967).
281. Mostyn, R. A. and Cunningham, A. F., *J. Inst. Petrol.* **53**, 101 (1967).
282. Mulford, C. E., *At. Absorption Newsletter* **5**, 28 (1966).
283. Nadirshaw, M. and Cornfield, A. H., *Analyst* **93**, 475 (1968).
284. Nakahara, T., Munemori, M. and Musha, S., *Anal. Chim. Acta* **50**, 51 (1970).
285. Olivier, M., *Z. Anal.-Chem.* **248**, 145 (1969).
286. Olson, A. D. and Hamlin, W. B., *At. Absorption Newsletter* **7**, 69 (1968).
287. Olson, A. D. and Hamlin, W. B., *Clin. Chem.* **15**, 438 (1969).
288. Omang, S. H., *Anal. Chim. Acta* **53**, 415 (1971).
289. Ottaway, J. M., Coker, D. T. and Davies, J. A., *Anal. Letters* **3**, 385 (1970).
290. Panday, V. K. and Ganguly, A. K., *Anal. Chim. Acta* **52**, 417 (1970).
291. Parker, J. E., *Spectrovision* **24**, 10 (1970).
292. Parker, M. W., Humoller, F. L. and Mahler, D. J., *Clin. Chem.* **13**, 40 (1967).
293. Perry, B., *Spectrovision* **25**, 8 (1971).
294. Peterson, E. A., *At. Absorption Newsletter* **8**, 53 (1969).
295. Pickles, D. and Washbrook, C. C., *Proc. Soc. Anal. Chem.* **7**, 13 (Jan. 1970).
296. Pinta, M., *Method. Phys. Anal.* (GAMS) **6**, 268 (1970).
297. Pitts, A. E. and Beamish, F. E., *Anal. Chim. Acta* **52**, 405 (1970).
298. Pitts, A. E., van Loon, J. C. and Beamish, F. E., *Anal. Chim. Acta* **50**, 181 (1970).
299. Pitts, A. E., van Loon, J. C. and Beamish, F. E., *Anal. Chim. Acta* **50**, 195 (1970).
300. Platte, J. A., in *Trace inorganics in water: Advances in Chemistry Series No.* 73, American Chemical Society, Washington D.C., 1968.
301. Platte, J. A. and Marcy, V. M., *At. Absorption Newsletter* **4**, 289 (1965).
302. Pollock, E. N., *At. Absorption Newsletter* **9**, 47 (1970).
303. Potter, A. L., Ducay, E. D. and McCready, R. M., *J. Assoc. Off. Anal. Chemists* **51**, 748 (1968).
304. Premi, P. R. and Cornfield, A. H., *Spectrovision* **19**, 15 (1968).
305. Price, W. J., *Effluent Water Treatment J.*, April 1967.
306. Price, W. J., *Proc. XIII Colloq. Intern. Spect.* Ottawa 1967, Adam Hilger, London, 1968, p. 291.
307. Price, W. J., *Paint, Oil, Colour J.* 3 (Aug. 21 1970).
308. Price, W. J. and Cooke. P. A., *Spectrovision* **16**, 7 (1966).
309. Price, W. J. and Cooke, P. A., *Spectrovision* **18**, 2 (1967).
310. Price, W. J. and Roos, J. T. H., *Analyst* **93**, 709 (1968).
311. Price, W. J. and Roos, J. T. H., *J. Sci. Food Agr.* **20**, 437 (1969).

312. Price, W. J., Roos, J. T. H. and Clay, A. F., *Analyst* **95**, 760 (1970).
313. Pybus, J., *Clin. Chim. Acta* **23**, 309 (1969).
314. Pybus, J. and Bowers, G. N., *Clin. Chem.* **16**, 139 (1970).
315. Pybus, J., Feldman, F. J. and Bowers, G. N., *Clin. Chem.* **16**, 998 (1970).
316. Ramakrishna, T. V., Robinson, J. W. and West, P. W., *Anal. Chim. Acta* **36**, 57 (1966).
317. Ramakrishna, T. V., Robinson, J. W. and West, P. W., *Anal. Chim. Acta* **45**, 43 (1969).
318. Ramakrishna, T. V., West, P. W. and Robinson, J. W., *Anal. Chim. Acta* **44**, 437 (1969).
319. Ramirez-Munoz, J., *Flame Notes, Beckman* **2**, 77 (1967).
320. Rawson, R. A. G., *Analyst* **91**, 630 (1966).
321. Beevers, J. R., *Econ. Geol.* **62**, 426 (1967).
322. Reid, J., Galloway, J. M., MacDonald, J. and Bach, B. B., *Metallurgia* **81**, 243 (1970).
323. Riley, J. P. and Taylor, D., *Anal. Chim. Acta* **40**, 479 (1968).
324. Riley, J. P. and Williams, A. P., *Mikrochim. Acta* **4**, 516 (1959).
325. Roach, A. G., Sanderson, P. and Williams, D. R., *Analyst* **63**, 42 (1968).
326. Robinson, J. W., *Anal. Chim. Acta* **23**, 458 (1960).
327. Robinson, J. W. and Hsu, C. J., *Anal. Chim. Acta* **43**, 109 (1968).
328. Rogers, G. R., *J. Assoc. Offic. Anal. Chemists* **51**, 1042 (1968).
329. Roos, J. T. H., *Spectrochim. Acta* **24B**, 255 (1969).
330. Roos, J. T. H. and Price, W. J., *Analyst* **94**, 89 (1969).
331. Roos, J. T. H. and Price, W. J., *J. Sci. Food Agr.* **21**, 51 (1970).
332. Roos, J. T. H. and Price, W. J., *Spectrochim. Acta* **26B**, 279 (1971).
333. Roos, J. T. H. and Price, W. J., *Spectrochim. Acta* **26B** 441 (1971).
334. Roosels, R. and Vanderkeel, J. V., *At. Absorption Newsletter* **7**, 9 (1968).
335. Rose, S. A. and Boltz, D. F., *Anal. Chim. Acta* **44**, 239 (1969).
336. Roussos, G. and Morrow, B., *Appl. Spectry*. **22**, 769 (1968).
337. Rubeska, I. and Moldan, B., *Appl. Opt.* **7**, 1341 (1968).
338. Rubeska, I. and Moldan, B., *Analyst* **93**, 148 (1968).
339. Rubeska, I. and Stupar, J., *At. Absorption Newsletter* **5**, 69 (1966).
340. Rubeska, I. and Svoboda, V., *Anal. Chim. Acta* **32**, 253 (1965).
341. Sachdev, S. L., Robinson, J. W. and West, P. W., *Anal. Chim. Acta* **37**, 12 (1967).
342. Saha, M. N. and Saha, H. K., *A Treatise on Modern Physics* Vol. 1, Allahabad, Calcutta, 1934.
343. Sandell, E. B., *Colorimetric Determination of Traces of Metals* Interscience, New York, 3rd ed. 1958.
344. Sanui, H. and Pace, N., *Appl. Spectry* **20**, 135 (1966).
345. Sastri, V. S., Chakrabarti, C. L. and Willis, D. E., *Can. J. Chem.* **47**, 587 (1969).
346. Sastri, V. S., Chakrabarti, C. L., and Willis, D. E., *Talanta* **16**, 1093 (1969).
347. Scarborough, J. M., *Anal. Chem.* **41**, 250 (1969).
348. Scarborough, J. M., Bingham, C. D. and DeVries, P. F., *Anal. Chem.* **39**, 1394 (1967).
349. Schallis, J. E. and Kahn, H. L., *At. Absorption Newsletter* **7**, 84 (1968).
350. Schnepfe, M. M. and Grimaldi, F. S., *Talanta* **16**, 1461 (1969).
351. Scholes, P. H., *Analyst* **93**, 197 (1968).
352. Scott, J. and Killer, F. C. A., *Proc. Soc. Anal. Chem.* **7**, 18 (Jan. 1970).
353. Selander, S. and Cramer, K., *Brit. J. Ind. Med.* **25**, 139 (1968).
354. Selander, S. and Cramer, K., *Brit. J. Ind. Med.* **25**, 209 (1968).
355. Shafto, R. G., *Prod. Finishing (Cincinnati)* **28**, 138 (1964).
356. Sheridan, J. E., *Spectrovision* **25**, 10 (1971).
357. Simon, R. K., Christian, G. D. and Purdy, W. C., *Amer. J. Clin. Pathol.* **49**, 207 (1968).
358. Simonian, J. V., *At. Absorption Newsletter* **7**, 63 (1968).
359. Simonson, A., *Anal. Chim. Acta* **49**, 368 (1970).
360. Singhal, K. C., Banerji, A. C. and Banerjee, B. K., *Technology (Sindri)* **5**, 117 (1968).
361. Slavin, S. and Slavin, W., *At. Absorption Newsletter* **5**, 106 (1966).
362. Slavin, W., *At. Absorption Newsletter* **4**, 243 (1965).
363. Smart, H. T. and Campbell, D. J., *Can. J. Pharm. Sci.* **4**, 73 (1969).
364. Smith, D. C., Johnson, J. R. and Soth, G. C., *Appl. Spectry* **24**, 576 (1970).

365. Smith, J. C. and Kench, J. E., *Brit. J. Ind. Med.* **14**, 240 (1957).
366. Smith, S. B., Blasi, J. A. and Feldman, F. J., *Anal. Chem.* **40**, 1525 (1968).
367. Soman, S. D., Panday, V. K. and Joseph, K. T., *Amer. Ind. Hyg. Assoc. J.* **30**, 527 (1969).
368. Spielholtz, G. I. and Toralballa, G. C., *Analyst* **94**, 1072 (1969).
369. Spitzer, H., *Z. Erzbergbau Metallhuettenw.* **19**, 567 (1966).
370. Sprague, S., Manning, D. C. and Slavin, W., *At. Absorption Newsletter* No. 20 p. 1 (May 1964).
371. Sprague, S. and Slavin W., *At. Absorption Newsletter* **3**, 160 (1964).
372. Sprague, S. and Slavin, W., *At. Absorption Newsletter* **4**, 367 (1965).
373. *Standard Methods of Testing Paint Varnish Lacquer and Related Products* H.M. Stationery Office, London, 1964.
374. Strasheim, A., Strelow, F. W. E. and Butler, L. R. P., *J.S. African Chem. Inst.* **13**, 73 (1960).
375. Strasheim, A. and Wessels. G. J., *Appl. Spectry*, **17**, 65 (1963).
376. Strelow, F. W. E., Feast, E. C., Mathews, P. M., Bothma, C. J. C. and Van Zyl, C.R., *Anal. Chem.* **38**, 115 (1966).
377. Strunk, D. H. and Andreasen, A. A., *J. Assoc. Offic. Anal. Chemists* **50**, 339 (1967).
378. Stupar, J. and Dawson, J. B., *Appl. Opt.* **7**, 1351 (1968).
379. Stupar, J. and Dawson, J. B., *At. Absorption Newsletter* **8**, 38 (1969).
380. Sullivan, J., Parker, M. and Carson, S. B., *J. Lab. Clin. Med.* **71**, 893 (1968).
381. Sullivan, J. V. and Walsh, A., *Spectrochim. Acta* **21**, 721 (1965).
382. Svec, H. J. and Anderson, A. R. Jnr., *Geochim. Cosmochim. Acta* **29**, 633 (1965).
383. Swider, R. T., *At. Absorption Newsletter* **7**, 111 (1968).
384. Sychra, V. and Matousek, J., *Anal. Chim. Acta* **52**, 376 (1970).
385. Sychra, V., Slevin, P. T., Matousek, J. and Bek, F., *Anal. Chim. Acta* **52**, 259 (1970).
386. Takeuchi, T. and Yanagisawa, M., *Japan Analyst* **15**, 1059 (1966).
387. Teclu, N. J., *J. Prakt. Chem.* **44**, 246 (1891).
388. Tenny, A. M., *Perkin-Elmer Instr. News* **18**, 1 and 14 (1967).
389. Thomerson, D. R., *Spectrovision* **25**, 12 (1971).
390. Thomerson, D. R. and Price, W. J., *Analyst* **96**, 321 (1971).
391. Thomerson, D. R. and Price, W. J., *Analyst* **96**, 825 (1971).
392. Thompson, A. J., *Spectrovision* **20**, 7 (1968).
394. Tindall, F. M., *At. Absorption Newsletter* **4**, 339 (1965).
395. Tindall, F. M., *At. Absorption Newsletter* **5**, 140 (1966).
396. Tolansky, S., *High Resolution Spectroscopy*, Methuen. London, 1947.
397. Tompsett, S. L., *Proc. SSOC. Clin. Biochem.* **5**, 125 (1968).
398. Torres, F., *Nat. Bur. Std. (U.S.) Reprint No.* SC-RR-69-784 Dec. 1969.
399. Toyoguchi, T. and Shimizu, H., *Japan Analyst* **16**, 565 (1967).
400. Trent, D. and Slavin, W., *At. Absorption Newsletter* No. 19, March 1964, p. 1.
401. Trent, D. J., *At. Absorption Newsletter* **4**, 348 (1965).
402. Trudeau, D. L. and Freier, E. F., *Clin. Chem.* **13**, 101 (1967).
403. Tyler, J. B., *At. Absorption Newsletter* **6**, 14 (1967).
404. Ulfvarson, U., *Acta Chem. Scand.* **21**, 641 (1967).
405. Ure, A. M. and Mitchell, R. L., *Spectrochim. Acta* **23B**, 79 (1967).
406. Uthe, J. F., Armstrong, F. A. J. and Stainton, M. P., *J. Fisheries Res. Board Can.* **27**, 805 (1970).
407. Vallee, B. L., *Clin. Chim. Acta* **25**, 307 (1969).
408. Van Assendelft. O. W., Zijlstra, W. G., Buursma, A., Van Kampen, E. J. and Hoek, W., *Clin. Chim. Acta* **22**, 281 (1968).
409. van Loon, J. C., *Z. Anal. Chem.* **246**, 122 (1969).
410. van Loon, J. C., Galbraith, J. H., Aarden, H. M., *Analyst* **96**, 47 (1971).
411. van Loon, J. C. and Parissis, C. M., *Analyst* **94**, 1057 (1969).
412. Veillon, C., Mansfield, J. M., Parsons, M. L. and Winefordner, J. D., *Anal. Chem.* **38**, 204 (1966).
413. Vink, J. J., *Analyst* **95**, 399 (1970).

414. Walker, C. R., Vita, O. A. and Sparks, R. W., *Anal. Chim. Acta* **47**, 1 (1969).
415. Walsh, A., *Spectrochim. Acta* **7**, 108 (1955).
416. Walsh, A., *Proc. XIII Colloq. Intern. Spect.*, Ottawa, Canada, 1967; Hilger, London, 1968, p. 257.
417. Walsh, A., *Atomic Absorption Spectroscopy* Plenary Lectures of International Conference, Sheffield, July 1969, IUPAC/Butterworths, London, 1970, p. 1.
418. Ward, G. M. and Miller, M. J., *Can. J. Plant Sci.* **49**, 53 (1969).
419. Watson, C. A., *Ammonium Pyrrolidine Dithiocarbamate*, Monograph 74, Hopkin and Williams, England.
420. Weger, S. J. Jr., Hossner, L. R. and Ferrar, L. W., *J. Agr. Food Chem.* **17**, 1276 (1969).
421. Weiner, J. P. and Taylor, L., *J. Inst. Brewing* **75**, 195 (1969).
422. Weir, D. R. and Kofluk, R. P., *At. Absorption Newsletter* **6**, 24 (1967).
423. Welcher, G. G., Kreige, O. H. and Owen, H., *At. Absorption Newsletter* **8**, 97 (1969).
424. Welcher, G. G. and Kriege, O. H., *At. Absorption Newsletter* **9**, 61 (1970).
425. Welz, B., S.A.C. *Anglo-Dutch Symposium on Detection of Major Components*, London, 1970.
426. Wendt, R. H. and Fassel, V. A., *Anal. Chem.* **38**, 337 (1966).
427. West, F. K., West, P. W. and Ramakrishna, T. V., *Environ. Sci. Technol.* **1**, 717 (1967).
428. West, T. S., *Atomic Absorption Spectroscopy* Plenary Lectures of International Conference, Sheffield, July 1969, IUPAC/Butterworths, London, 1970, p. 99.
429. West, T. S. and Williams, X. K., *Anal. Chem.* **40**, 335 (1968).
430. West, T. S. and Williams, X. K., *Anal. Chim. Acta* **42**, 29 (1968).
431. West, T. S. and Williams, X. K., *Anal. Chim. Acta* **45**, 27 (1969).
432. White, R. A., *Intern. At. Abs. Conf.*, Sheffield, 1969, Abstract G5.
433. Whittington, C. M. and Willis, J. B., *Plating* **51**, 767 (1964).
434. Williams, D. R., *Spectrovision* **19**, 8 (1968).
435. Williams, C. H., David, D. J. and Iismaa, O., *J. Agr. Sci.* **59**, 381 (1962).
436. Willis, J. B., *Nature* **184**, 186 (1959).
437. Willis, J. B., *Spectrochim. Acta* **16**, 273 (1960).
438. Willis, J. B., *Nature* **186**, 249 (1960).
439. Willis, J. B., *Spectrochim. Acta* **16**, 259 (1960).
440. Willis, J. B., *Spectrochim. Acta* **16**, 551 (1960).
441. Willis, J. B., *Nature* **191**, 381 (1961).
442. Willis, J. B., *Anal. Chem.* **33**, 556 (1961).
443. Willis, J. B., *Anal. Chem.* **34**, 614 (1962).
444. Willis, J. B., *Methods of Biochemical Analysis*, Interscience, New York, 1963, Vol. XI, p. 1.
445. Willis, J. B., *Nature* **207**, 715 (1965).
446. Wilson, L., *Anal. Chim. Acta* **30**, 377 (1964).
447. Wilson, L., *Metallurgy Note* 42, Aeronautical Research Laboratories, Dept. of Supply, Melbourne, Australia, 1966.
448. Wilson, L., *Anal. Chim. Acta* **40**, 503 (1968).
449. Wilson, A. L., *Chem. Ind.* **36**, 1253 (1969).
450. Winefordner, J. D. *Atomic Absorption Spectroscopy* Plenary Lectures of International Conference, Sheffield, July 1969, IUPAC/Butterworths, London, 1970, p. 35.
451. Winefordner, J. D. and Staab, R. A., *Anal. Chem.* **36**, 1367 (1964).
452. Winefordner, J. D. and Vickers, T. J., *Anal. Chem.* **36**, 161 (1964).
453. Wollaston, W. H., *Phil. Trans. Roy. Soc. London Ser. A* **92**, 365 (1802).
454. Woolley, J. F., *Spectrovision* **22**, 7 (1969).
455. Yamamoto, Y., Kumamaru, T., Hayashi, Y. and Otani, Y., *Japan Analyst* **17**, 92 (1968).
456. Yamamoto, Y., Kumamaru, T., Hayashi, Y. and Otani, Y., *Anal. Letters* **1**, 955 (1968).
457. Yanagisawa, M., Suzuki, M. and Takeuchi, T., *Anal. Chim. Acta* **46**, 152 (1969).
458. Zaugg, W. S. and Knox, R. J., *Anal. Chem.* **38**, 1759 (1966).
459. Zeeman, P. B. and Butler, L. R. P., *Appl. Spectry.* **16**, 120 (1962).
460. Zettner, A. and Mensch, A. H., *Amer. J. Clin. Pathol.* **49**, 196 (1968).

461. Zettner, A. and Seligson, D., *Clin. Chem.* **10**, 869 (1964).
462. Zettner, A., Sylvia, L. C. and Capacho-Delgado, L., *Amer. J. Clin. Pathol.* **45**, 533 (1966).
463. Zlatkis, A., Bruening, W. and Bayer, E., *Anal. Chem.* **41**, 1692 (1969).
464. Zurlo, N., Griffini, A. M. and Colombo, G., *Anal. Chim. Acta* **47**, 203 (1969).
465. Aldous, K. M., Bailey, B. W. and Rankin, J. M., *Anal. Chem.* 1972, **44**, 191.
466. Bratzel, M. P., Dagnall, R. M. and Winefordner, J. D., *Anal. Chem.* 1969, **41**, 1527.
467. Quarrell, T. M., Powell, R. J. W. and Cluley, H. J., *Analyst* 1973, **98**, 443.
468. Thomerson, D. R., *Spectrovision* 1972, **26**, 13.
469. Willis, J. B. *Applied Optics* 1968, **7**, 1295; see also chapter in *Analytical Flame Spectroscopy* (Ed. R. Mavrodineanu), Macmillan, London, 1970.

Index